现代物理导论

INTRODUCTION TO
MODERN PHYSICS

王 一 编著

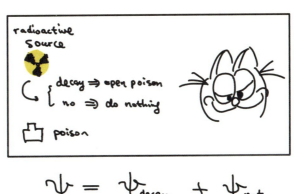

中国科学技术大学出版社

内容简介

本书挑选 20 世纪物理学发展中最激动人心的内容进行介绍。主要包括：相对论与宇宙学，重点讲述狭义相对论，简述广义相对论和宇宙学；量子力学与原子，重点讲述初步的量子力学，简述量子信息和原子物理；简单与复杂，简述作用量、对称性、守恒律、粒子与弦、熵和复杂系统。

本书适合物理系和其他理工类专业具有普通物理、微积分基础的读者使用。内容也分别对应 Coursera 在线教育平台上的三门慕课（MOOC）课程，可以配合教学视频使用。

图书在版编目(CIP)数据

现代物理导论 ＝ Introduction to Modern Physics：英文／王一编著. -- 合肥：中国科学技术大学出版社，2024．10. -- ISBN 978-7-312-05967-4

Ⅰ．O4

中国国家版本馆 CIP 数据核字第 202479CN47 号

现代物理导论

XIANDAI WULI DAOLUN

出版	中国科学技术大学出版社 安徽省合肥市金寨路 96 号，230026 http://press.ustc.edu.cn https://zgkxjsdxcbs.tmall.com
印刷	安徽国文彩印有限公司
发行	中国科学技术大学出版社
开本	787 mm×1092 mm　1/16
印张	16
字数	440 千
版次	2024 年 10 月第 1 版
印次	2024 年 10 月第 1 次印刷
定价	80.00 元

Preface

The 20th century is often hailed as the "century of physics". In this century, our understanding of space, time and fundamental structure of matter are completely reformed. This book aims to showcase the captivating aspects of physics developed over the past a hundred years. As a theoretical physicist, my focus lies in elucidating the underlying structure of various physical theories and showing the profound insights that physicists have gained about Nature. This book is not intended as popular science, or a collection or dictionary-like explanations of concepts, but rather attempts to show in some depth on the topics that covered, uncovering the intrinsic surprises and beauty brought by physics.

The intended readers of this book are those who have already learned calculus and general physics (especially mechanics). With this necessary background, the book is designed to cater to two types of readers: (1) physics majors seeking a broad overview of the subject to facilitate more focused and purposeful future studies, and (2) college students from other disciplines, such as mathematics and engineering, who may not continue their physics education beyond general physics. For the latter group, this book serves as a bridge to the fascinating world of modern physics. This book may be used as a textbook of the second year university students at schools of physics and engineering for this purpose.

This book is organized into three main parts: (1) Relativity and Cosmology, (2) Quantum Physics, and (3) Simplicity and Complexity.

The Relativity and Cosmology section covers Special Relativity, General Relativity, and Cosmology.

The Quantum Physics section delves into Quantum Mechanics, Atomic Physics, and Quantum Information.

The Simplicity and Complexity section explores the Principle of Least Action, Particle

Physics, Entropy and Information, and Complex Systems.

Each of the aforementioned topics is vast enough to warrant its own textbook, and all such classics already exist. Given the goals and scope of this book, it is not possible to cover each topic exhaustively. Instead, we aim to present a comprehensive introduction to the basic ideas and fundamental frameworks of the two major 20th-century breakthroughs: Relativity and Quantum Mechanics, and based on that, provide a guided tour of the century's most influential physics developments.

While we strive to offer a broad picture of 20th-century physics, it is important to note that some areas have been omitted or only briefly touched upon, such as condensed matter physics (which needs a deeper quantum mechanics background to get into in depth), laser physics and material science.

This book is complemented by video materials available through three MOOC courses on Coursera: "Understanding Modern Physics Ⅰ: Relativity and Cosmology", "Understanding Modern Physics Ⅱ: Quantum Mechanics and Atoms", and "Understanding Modern Physics Ⅲ: Simplicity and Complexity". These videos can aid readers in grasping the concepts presented in the book. Additionally, the popular science book *Yi on theories of everything* (一说万物, in Chinese) co-launched by the author and the Micius Salon (墨子沙龙), covers similar physics content in the depth of popular science and can be considered a popular science companion of this book. Readers who find this book challenging may benefit from reading it alongside with *Yi on theories of everything*.

This book is based on the author's Modern Physics course taught at the Hong Kong University of Science and Technology (HKUST). I would like to express my gratitude to HKUST, and especially its Department of Physics, for their support in my teaching endeavors. I would like to thank my teaching assistants of this course over the years, especially Zhou Siyi, Tong Xi, Wang Zun and Li Cheungshun for their help. I humbly acknowledge my own limitations and wholeheartedly welcome feedback and corrections from readers.

We have used the following abbreviations to save some writing:

wrt: with respect to

LHS: the left hand side (of an equation)

RHS: the right hand side (of an equation)

EoM: the equation of motion

ODE: ordinary differential equation

PDE: partial differential equation

GR: general relativity

Contents

Preface i

Chapter 1 Special Relativity 1

   1.1 Principles of Special Relativity 2
      1.1.1 Galileo's Principle of Relativity 2
      1.1.2 The Speed of Light 4
      1.1.3 Einstein's Relativity 10
   1.2 Time Dilation 13
   1.3 Physical Picture and Physical Intuition 20
   1.4 Length Contraction 23
   1.5 Meaning of the "Same Time" (Simultaneity) 27
      1.5.1 Simultaneity Depends on Which Observer 27
      1.5.2 A Spacetime Diagram of the Above Scenario 28
      1.5.3 Causality and Types of Separations 32
   1.6 Example: The Ladder Paradox 35
   1.7 The Lorentz Transformation 39
   1.8 The Geometry of Spacetime 44
   1.9 Relativistic Momentum and Energy 51
      1.9.1 Relativistic Momentum 52
      1.9.2 Relativistic Energy 56
   1.10 Epilogue: Summary and What's Next 60

Chapter 2 General Relativity 68

   2.1 The Equivalence Principle 68

2.2	Time with Uniform Gravity	72
2.3	Black Holes	75
2.4	Gravitational Waves	77
2.5	Epilogue: Summary and What's Next	80

Chapter 3 Cosmology 83
3.1	The Dynamics of the Universe	84
3.2	The Early Universe	89
3.3	Epilogue: Summary and What's Next	91

Chapter 4 Quantum Mechanics 93
4.1	The Nature of Light	94
	4.1.1 Is Light Particles or Waves?	94
	4.1.2 The Photoelectric Effect	97
	4.1.3 The Single Photon Double-Slit Experiment	103
	4.1.4 A Wave-Particle Duality, and from Light to All Matter	106
4.2	The Quantum Wave Function	108
	4.2.1 The Wave Function as a Probability Amplitude	109
	4.2.2 Consequence of Superposition and Linearity	111
	4.2.3 Extracting Momentum Information from the Wave Function	113
4.3	Observables and Measurements	115
4.4	The Uncertainty Principle	118
4.5	The Schrödinger Equation	122
	4.5.1 The Schrödinger Equation	122
	4.5.2 A Step in the Potential	125
	4.5.3 The Potential Barrier: Reflection and Tunneling	128
	4.5.4 The Potential Well: Scattering and Bound States	130
4.6	Identical Particles	135
4.7	Epilogue: Summary and What's Next	137

Chapter 5 Atoms 142
5.1	How Did We Know That Matter Is Made of Atoms?	143
5.2	The Hydrogen Atom	150
	5.2.1 Aspects of Observations	150
	5.2.2 Bohr's Model of the Hydrogen Atom	152
	5.2.3 (Optional)The Schrödinger Equation of the Hydrogen Atom	153
	5.2.4 Properties of the Hydrogen Atom	157
5.3	The Periodic Table	159
5.4	Epilogue: Summary and What's Next	162

Chapter 6 Entanglement and Quantum Information 164
6.1 Spin 165
6.1.1 The Stern-Gerlach Experiment 165
6.1.2 The Quantum Mechanics of Spins 169
6.2 Multiple Spins and Their Entanglement 173
6.2.1 Multiple Spins, Entanglements, and Quantum Computing 173
6.2.2 The Einstein-Podolsky-Rosen (EPR) Paradox 176
6.2.3 The Bell's Inequality and Its Violation 179
6.3 Epilogue: Summary and What's Next 181

Chapter 7 From the Action to the Laws of Nature 183
7.1 Fermat's Principle of Light 184
7.2 Principle of Extremal Action 188
7.3 Symmetry and Conservation Laws 192
7.4 The Hidden Quantum Reality 197
7.5 Epilogue: Summary and What's Next 199

Chapter 8 From Particles to Strings 202
8.1 Elementary Particles 202
8.2 Quantum Gravity 206
8.2.1 Do We Need Quantum Gravity, and Where to Find It? 206
8.2.2 Theoretical Challenges 209
8.2.3 Experimental Challenges 212
8.3 Is the World Made of Strings? 215
8.4 Epilogue: Summary and What's Next 219

Chapter 9 Entropy and Information 220
9.1 The Statistical Entropy 221
9.2 The Arrow of Time 223
9.3 Entropy and Information 226
9.4 Epilogue: Summary and What's Next 229

Chapter 10 Complexity 231
10.1 Iteration: from Population of Rabbits to Chaos 232
10.2 Fractals: Dimensions Reloaded 236
10.3 Epilogue: Summary and What's Next 241

Index 242

Chapter 1
Special Relativity

Let us start our journey by a thought experiment (Fig. 1.1)—Alice is in a spaceship to a star, 5 light years away from us, and will return to the earth immediately after her arrival. Bob sees her off and waits her to come back. Suppose the spaceship runs almost as fast as the speed of light c, say, $v = 0.995c$.

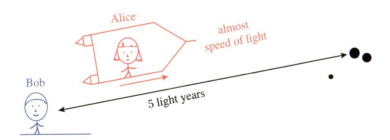

Fig. 1.1 Alice travels to a star by a spaceship

Amazingly, at $v \sim c$, things behave dramatically different from our daily experience (Newtonian mechanics). Let's see a few surprising facts that they find:

Bob's observations about Alice's journey

(1) Bob wrote a letter to Alice, and incidentally Alice wrote a letter to Bob around the same time. Bob thinks that he wrote his letter earlier, but Alice insists that she wrote her letter earlier. They have considered that light needs time to travel. But they cannot resolve the dispute even after that.

(2) Bob finds the spaceship much "heavier"— when $v \sim c$, Alice needs increasingly greater amount of energy to accelerate the spaceship even by a little.

(3) When Alice comes back, 10 years has passed for Bob. But for Alice, her clock, her feeling, everything about Alice indicates that only 1 year has passed.

What happened? After learning this part, you will find it out, and much more — In fact, we have to think about space and time in a totally different way from we have naively thought.

Understanding spacetime

We find deeper understanding of spacetime as we understand nature better. The efforts have not come to an end so far.

Newtonian: space and time are absolute and independent "playgrounds" for matter.

Special relativity: space and time are unified, and depends on motion of observers (like relative orientation of an object depends on rotation of observers).

General relativity: space and time can be curved by objects.

Quantum gravity (conjectured): space and time may be emergent from holography, quantum entanglements ... We have not fully understood it yet, but the hints are profound.

1.1 Principles of Special Relativity

1.1.1 Galileo's Principle of Relativity

Imagine: As in Fig. 1.2, Alice is moving with constant velocity v in a closed car with respect to (wrt for short) the ground. If Alice does not look outside the car, how can Alice find out that she is moving wrt the ground?

No. Whatever activities/experiments Alice tries, she finds no difference from if the car is not moving. To put it in the language of physics, the laws of nature she probes is

---

① By "look outside", we mean connections by any means to the outside of the car, including using light, sound, gravitational waves, etc. We neglect rotation of the earth such that Alice and Bob are in inertial frames (constant velocity).

identical to the laws of nature probed when she is not moving.

In 1632, Galileo asserted that this was true for all physical laws, and for all inertial frames. This is known as Galileo's principle of relativity. It is impressive to note that this nearly 400-year-old principle still holds now to the best of our knowledge.

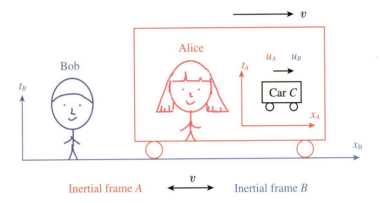

Fig. 1.2   Alice has her time t_A and space x_A; Bob has his time x_B and coordinate x_B. Alice and Bob have relative speed v (because Alice is standing on a moving big car). A small car C has speed u_A wrt Alice and u_B wrt Bob. For simplicity, only one space direction x is shown and other space dimensions y and z are suppressed.

Galileo's principle of relativity:

Laws of nature take the same form in all inertial frames.

Let us refer to Galileo's principle of relativity by ($\mathbb{R}$) as we will use it many times. There are some alternative equivalent statements of ($\mathbb{R}$):

(1) Motion is relative.

(2) There is no absolute sense of "who moves".

The change (by a relative velocity) from one inertial frame to another is called a "boost". Thus we can also say

(3) Laws of nature are not changed by a boost.

> ### Some Histories of Galileo's Principle of Relativity

Nowadays, ($\mathbb{R}$) looks pretty trivial. This is because you have already learned Newtonian mechanics.

But at Galileo's (1564—1642) time, people had no concept of acceleration; people relied on naked eyes to do astronomy. Especially, people believed in the geocentric model more than Helio-centrism (earth moves around the sun). An argument for geocentric model was that we would have fallen out of the earth if the earth is moving fast. (ℝ) shows that it will not happen.

Moreover, now we know that (ℝ) still holds in situations where Newtonian mechanics does not hold, including special relativity and beyond.

Newtonian mechanics is consistent with Galileo's relativity

In Fig. 1.2, does the Newtonian 2nd law (the law of nature) take the same form wrt Alice and Bob? To check that, note the relation between their reference frames are

$$t_B = t_A, \qquad x_B = x_A + v t_A \tag{1.1}$$

The 2nd law in Alice's frame is $F = ma_A$. What about in Bob's frame? Bob picks up Alice's equation and transform it into that in Bob's form:

$$F = ma_A = m\ddot{x}_A = m\ddot{x}_B = ma_B \tag{1.2}$$

Thus Bob indeed has the same 2nd law in his frame.

As a reminder for later reference, the velocity addition rule in Newtonian mechanics can be derived from the same frame transformation (1.1):

$$u_A = \dot{x}_A, \qquad u_B = \dot{x}_B = u_A + v \tag{1.3}$$

From the above, we observe: (ℝ) is respected by Newtonian mechanics. However, Newtonian mechanics is not the only system that satisfies (ℝ). To see that, and to introduce Einstein's relativity, let us introduce another fundamental principle of nature.

1.1.2 The Speed of Light

When you turn on the light, light "immediately" fills the room. In our everyday life, the speed of light is so fast that we usually ignore the propagation time needed for light. However, is the speed of light infinite, or finite?

---

① We use the notation $\dot{x} \equiv \mathrm{d}x/\mathrm{d}t$, $\ddot{x} \equiv \mathrm{d}^2 x/\mathrm{d}t^2$. Can you also verify Newtonian 1st and 3rd laws?

(Optional) Finite c observed

As early as in 1676, Rømer noted that the actually observed eclipses of the Jupiter's moon "Io" and the calculated time had a difference, when the eclipses happened with earth position L, K, G and F (Fig. 1.3). This is interpreted as: light needs time to travel through intervals LK and GF.

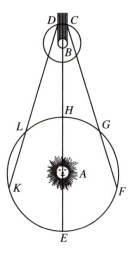

Fig. 1.3

Many delicate experiments have been designed to show that, the speed of light is actually finite at $c \approx 3 \times 10^8$ m/s. Nowadays, the finiteness of speed of light is not only a reality, but also has great potential for future applications. For example, one can search videos for "seeing what's behind a wall through reflection" for applications for the finite speed of light.

The exact value of the speed of light

The speed of light c has an exact value: $c = 299792458$ m/s.

Are you surprised by this fact? Usually, physical constants are determined by measurements. Measurements always have errors. Then, how can c have an exact value?

The modern definition of a meter (since 1983) is (distance travelled by light in 1 second)/(value of c). Indeed, historically (1889—1960) meter was defined by a real object known as "International Prototype Meter". But defining meter using a real object has many disadvantages, including

(1) One has to physically compare with the prototype (or its copies) to determine

length.

(2) The accuracy of the copies made are limited by the technology of the time.

(3) The length of the prototype change slightly over time. And it may be damaged.

By defining meter with a constant of nature c, anyone can reproduce the meter definition. The error is only limited by his/her precision of experiments. Also, the modern definition of time and mass use the fundamental natures of quantum mechanics, through atomic frequencies and the Planck constant.

Now, let's come to the key of the section, and the corner stone of Einstein's special relativity. ①

What's the speed of light from a moving source?

Bob holds a candle (Fig. 1.4). The speed of light from Bob's candle is $c_B = c$ wrt Bob. Alice is moving at speed v wrt Bob in a closed car. She also holds a candle. For the light from Alice's candle:

(1) What is the speed c_A wrt Alice?

(2) What is the speed c_B wrt Bob?

Fig. 1.4

We can immediately get $c_A = c$ from (ℝ). Because otherwise Alice would know that she is moving by noticing a different value of speed of light without interacting with the outside of the car. However, what is c_B for the light from Alice's candle?

Naively, we would have expected that $c_B = c_A + v = c + v$ from the Newtonian speed addition rule (1.3). And this looks natural — in our everyday lives (speeds much slower than light): if Alice is in a car moving with $v \ll c$ (wrt Bob), and throw a ball with speed u_A wrt Alice, the speed of the ball wrt Bob should be $u_B = v + u_A$.

However, the light propagation should be computed by Maxwell equations. As a result of computation, one obtains $c_B = c$! We will below derive this surprising fact for you. Multi-variable calculus is used, which is beyond the prerequisite math. Thus if you

---

① We have said that the modern speed of light is a way to define length. But from now on, let's go back to the early 1900s, and still consider the speed of light as measured by distance moved per unit time, with common sense definition of distance and time duration.

cannot follow the derivation, it's okay to skip it.

> **Maxwell and relativity**
>
> In fact, hidden in Maxwell's equations, there is a symmetry, which already implies Einstein's special relativity. The invariant speed of light is just a consequence of that. The symmetry is known as Lorentz transformations, which is developed by Voigt, Lorentz, Larmor and Poincare during 1887—1905 before Einstein established special relativity. We will come to the Lorentz transformations in Section 1.7 from a mechanical point of view.

(Optional) Speed of light from Maxwell equations

Let us work in Bob's reference frame. The Maxwell's equations with vacuum permeability μ_0 and permittivity ϵ_0 are:

$$\nabla \cdot \boldsymbol{E} = \frac{\rho}{\epsilon_0} \tag{1.4}$$

$$\nabla \cdot \boldsymbol{B} = 0 \tag{1.5}$$

$$\nabla \times \boldsymbol{E} = -\frac{\partial \boldsymbol{B}}{\partial t} \tag{1.6}$$

$$\nabla \times \boldsymbol{B} = \mu_0 \left[\boldsymbol{J} + \epsilon_0 \frac{\partial \boldsymbol{E}}{\partial t} \right] \tag{1.7}$$

As we study the propagation of light, the Maxwell equations are used in the vacuum environment. Thus the charge density $\rho = 0$, and the current $\boldsymbol{J} = 0$. To eliminate $\boldsymbol{E}$ and obtain an equation for $\boldsymbol{B}$ only, we use the following trick: Let us apply $\nabla \times (...)$ to the LHS and RHS of Eq.(1.7), respectively:

$$\nabla \times \text{LHS} = \nabla \times (\nabla \times \boldsymbol{B}) \xrightarrow{\text{math identity}} \nabla(\nabla \cdot \boldsymbol{B}) - \nabla^2 \boldsymbol{B} \xrightarrow{\text{using Eq. (1.5)}} -\nabla^2 \boldsymbol{B}$$

$$\nabla \times \text{RHS} = \mu_0 \epsilon_0 \frac{\partial}{\partial t}(\nabla \times \boldsymbol{E}) \xrightarrow{\text{using Eq. (1.6)}} -\mu_0 \epsilon_0 \frac{\partial^2}{\partial t^2} \boldsymbol{B}$$

Thus, LHS=RHS is a wave equation:

$$\frac{\partial^2}{\partial t^2} \boldsymbol{B} - \frac{1}{\mu_0 \epsilon_0} \nabla^2 \boldsymbol{B} = 0 \tag{1.8}$$

In a course of mathematical method of physics, one will study how to systematically solve this equation. Here we will not do so. Instead, we can always propose a solution and check it indeed solves Eq. (1.8) without using more math. One can check that the below E&M wave is a solution:

$$B_z = B_0 \cos[k(x - ct)], \qquad \text{where } c \equiv \frac{1}{\sqrt{\mu_0 \epsilon_0}} \tag{1.9}$$

The take home message from the calculation is: once a beam of E&M wave is emitted, it can then propagate in the vacuum without the reference of the emitter, i.e. the motion information of the emitter is "forgotten". The speed of light is calculated by constants of nature μ_0 and ϵ_0, independent on the speed of the emitter (or the observer). Thus we conclude that the answer is $c_B = c$, the same speed that Alice observes (i.e. $c_B = c_A$)!

Velocity from phase

One can read off velocity (more accurately: phase velocity) from the "phase" factor $\cos[k(x - ct)]$ in Eq.(1.9). This can be done by following one period of oscillation in the below figure and see how it moves.

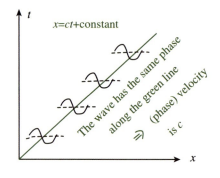

Fig. 1.5

We will encounter similar wave equations in quantum mechanics later in this course, and use complex solutions $\exp[ik(x - ct)]$. Thus, make sure to understand why (1.9) describes a moving wave. If you think at a deeper level, using phase velocity above is in fact not the way to describe how fast information propagates. Then you need to solve the Maxwell equations more carefully, and find a regarded solution produced by accelerating source. It is beyond the scope of the current discussion.

There must be something terribly wrong at this moment: The velocity addition rule (1.3) is inconsistent with the observer-independent speed of light (1.9). In such a situation, we need input from experiments to see who is right. The Michelson-Morley interferometer experiment (1887) shows that Maxwell is right. The Newtonian velocity addition does not apply to light!

> **Modern interferometers**
>
> The Michelson-Morley interferometer was designed to measure the variation of speed of light. Now that we know (at least assume) that the speed of light is truly a constant, we can turn the same interferometer into a precise measurement of the variation of length of the two arms. This has many applications (check Wikipedia for example), from performing Fourier transformation to measuring the diameter of stars. Last but not least, this is also how LIGO detected gravitational waves in 2015/2016.

The Michelson-Morley interferometer experiment

See Fig. 1.6, a beam of light is emitted from the source. Via a half-silvered mirror, the beam is split into two, reflected by two mirrors respectively and re-combine on the screen. Due to different length of propagation, interference fringes develop.

Note that the experiment is done on the earth. The earth moves at $v \approx 30$ km/s around the sun. Thus if the velocity addition rule applies, the speed of light in the Michelson-Morley interferometer should change under rotation.

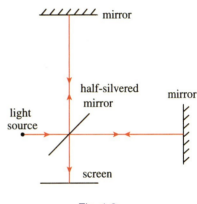

Fig. 1.6

Then the interference fringes should shift. But the shift is not observed, indicating no velocity addition. Maxwell wins over Newton.

> **Is Newton "wrong"?**
>
> Now that the velocity addition (in the core of Newtonian mechanics) fails for light, is Newton wrong? Should we abandon Newtonian mechanics?
>
> In modern physics, we think of most (perhaps all) theories as "effective" theories — they apply with certain approximations. When $v \ll c$, Newtonian mechanics is still useful, as it is pretty precise, and simpler and more intuitive than special relativity.
>
> Thus, by learning special relativity, we do not mean to abandon Newtonian mechanics. Rather, Newtonian mechanics is still much more widely used than special relativity in the modern era (since we do not usually move at $v \sim c$ or care about the corrections suppressed by v/c).

In fact, special relativity has a similar fate — it is classical. When considering small enough particles, one has to consider quantum mechanics and we arrive at quantum field theory. And considering the full quantum effects of gravity, quantum field theory is again not adequate...

(Optional) Historical remarks: the story of aether

In fact, what we have introduced here is *not* what researchers were thinking 100 years ago when they made the discoveries.

At the time of Maxwell, it was believed that E&M waves propagates in a media called aether, analogous to sound waves propagating in the air. Thus, the speed of light $c = 1/\sqrt{\mu_0 \epsilon_0}$ is wrt aether as well. The parameters μ_0 and ϵ_0 were considered the property of aether, instead of the property of the vacuum.

Then, the velocity addition rule had no problem — just as velocity addition of wind and sound waves propagating in the wind.

Now the question is: when the earth orbits the sun, do aether around the earth move together with the earth, or remain static?

The observation of stellar aberration (1600s—1700s) showed that aether did not move together with the earth. And in the context of aether, the Michelson-Morley experiment showed that aether indeed moved together with the earth. This was the contradiction that puzzled physicists 100 years ago. And Einstein's contribution is to eliminate the need of aether with his theory of relativity.

As aether is abandoned, now when we learn E&M waves, we do not introduce aether any more. Thus, I choose to use the way of introduction closer to modern thinking instead of historical thinking, only leaving a remark of aether within this historical remark.

1.1.3 Einstein's Relativity

How to resolve the contradiction between Maxwell's observer-independent c and the velocity addition rule? This is one of the two biggest problems in physics at the beginning of the 20th century.

When Einstein was 16 years old, he was already deeply puzzled by light: What if I run as fast as light? Will light be at rest? 10 years later, he found out the solution.

Here is how Einstein solved the problem in 1905: he took the observer-independency of c as a fundamental assumption of his theory. It is assumed, so problem solved.

Seriously, an assumption is no more than an assumption. Everyone can make assumptions. But what's truly revolutionary is the implications of the assumption and how to get a consistent theory on top of that. To be clear, here we summarize the assumptions, being prepared to proceed to the profound implications.

> **Remarks about the speed of light**
>
> The speed of light is observer-independent, but not velocity. In other words, the direction of light can be observer-dependent. We will see this explicitly later when we construct a "light clock".
>
> The frequency of light is observer dependent (relativistic Doppler effect). When you move towards light, the light frequency becomes higher (blue shift). When you move away from light, the light frequency becomes lower (redshift).
>
> In non-vacuum situations, the speed of light will depend on the motion of the media. The speed of light in the moving frame of media can be obtained using relativistic velocity addition rules to be discussed later.

The postulates of special relativity

($\mathbb{R}$) The relativity principle: Laws of nature keep the same form in all inertial frames.
($\mathbb{C}$) The vacuum speed of light is c in all inertial frames.

The postulates ($\mathbb{R}$) and ($\mathbb{C}$) are the original postulates in special relativity, and often referred to as the only assumptions. But to be clear, one has also to postulate observer-independency of events to complete the theoretical basis of special relativity.

The postulate of observer-independent events

($\mathbb{E}$) Let us define an event to be something that happened to a particular small object at a particular moment (i.e. *localized* at a spacetime point). Then, the occurrence (or not) of an event is observer-independent. If one observer finds that an event E happened, all observers who saw the event must agree, no matter how they move.

This postulate looks too trivial to mention. Nevertheless, let us make things clear.

For example, the followings are events that all (honest) observers must agree:

(1) "A beam of light is reflected by a mirror."

(2) "Alice and Bob met. At the time when they met, their watches both pointed to 10 am."

However, the following is NOT an event:

(3) "Alice and Bob are 5 light years away. When Alice's watch pointed to 10 am, Bob's watch also pointed to 10 am."

This is not an event because Alice and Bob are at different locations. So it may be possible that an observer Charlie considers the above statement to be true; while his running Dog considers the above statement to be false.

> **Not one, but two events**
>
> In the example of 5-light-year-separated Alice and Bob, the sentence is not an event, but one can separate it into two events: When Alice looked at her watch, her watch shows 10 am. And when Bob looked at his watch, his watch shows 10 am. The same applies for the comparison of ruler example. It makes sense to talk about relations of events in relativity, as we will see later.

Similarly, the following is NOT an event and may not be agreed by all observers:

(4) "A moving ruler has the same length with a static ruler because the two end points coincided at the same moment."

Remark: the time of events wrt a reference frame

With finite speed of light, we need a clarification about the time of events. For example, an event took place some distance away from Bob. When talking about the time of the event, it could refer to:

The time in Bob's reference frame. Imagine Bob puts a clock at every spatial point[1] in his frame, which "records" the time that an event happened.

The time that Bob actually "see" the event. This needs to take the time of light propagation between the event and Bob. This is trivial but messy.

---

[1] By "a clock at every spatial point": the clocks are static wrt Bob, and synchronized wrt Bob's frame (the coordinate time in his coordinate system). Practical synchronization can be done by using light signals to compare clocks, and deduct the time used for light propagation.

By "record": imagine that Bob has a helper at every point in his frame (i.e. coordinate system). Once any event happened, the helper can immediately write down Bob's time and spatial coordinate of the event. This will be denoted by the coordinate of the event (t_B, x_B) wrt Bob.

In this course, unless otherwise emphasized, we will by default use the time in Bob's frame, instead of the time that the Bob actually see the light signal of the event reaching him, when talking about the time wrt Bob.

With the three postulates, we are now prepared to start the adventure of special relativity.

1.2 Time Dilation

At the beginning of this part, we promise to explain why the space traveler Alice is younger than Bob when she returns (if she initially has the same age as Bob before space travel). This is the goal of this section.

To do this, we have to touch the most fundamental concept in physics—time. Let us begin by recalling how Newton commented about time.

Time in Newton's Principia

"Absolute, true and mathematical time, of itself, and from its own nature flows equably without regard to anything external, and by another name is called duration; relative, apparent and common time, is some sensible and external (whether accurate or unequable) measure of duration by the means of motion, which is commonly used instead of true time."
— Newton, *Philosophiæ Naturalis Principia Mathematica*

Thus, Newton believed that the "true time" exists independent of anything else. If so, can anything access and probe the true time?

(1) If it cannot be accessed, why bother to define "true time"? Why shouldn't we shave this concept using Occam's razor?

(2) If it can be accessed by an apparatus, why it cannot be altered? As an extrapolation of the Newtonian's third law, when there is some action from the true time to the apparatus, why does the apparatus not back-react on the true time and alter it?

Fine. We are physicists. What can we do to get rid of the metaphysical thinking about time? Let us do a practical thing — define time by actually making a simplest clock: the light clock.

> **What can we trust?**
>
> As now we are to go beyond Newton, you may wonder, what can we trust. Of course we want to stand on Newton's shoulder and make use of some of his legacy.
>
> Almost all results for low speed motion ($v \ll c$) can be trusted (with the exception of rest energy, which we will come back to later). For high speed motion ($v \sim c$), we trust ($\mathbb{R}$), ($\mathbb{C}$), ($\mathbb{E}$) and nothing else.
>
> For example, when discusing time: we trust that Alice's mechanical watch (static wrt Alice) defines time nicely for Alice. Because the time from mechanical watch is as a result of low speed mechanical motion. But we cannot trust Newton's conclusion about how Alice's time (as defined by the tick-tock of her mechanical watch) lapse wrt Bob, if Bob has relative motion wrt Alice with $v \sim c$. When discussing length: we trust that Alice's ruler (static wrt Alice) defines length for Alice but we cannot trust how Bob observes the length of Alice's ruler.

Shut up and construct a light clock

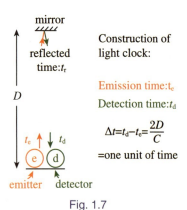

Fig. 1.7

We define time by a tick-tock of the "light clock" (Fig. 1.7). The tick-tock time interval Δt is the time between the emitter emits the light (at time t_e) and the detector detects the light (at time t_d). In the middle, the light is reflected by a mirror (at time t_r). With the light clock, a "standard time interval" can be defined as [①]

$$\Delta t \equiv t_d - t_e = \frac{2D}{c} \qquad (1.10)$$

Then time can be measured by number of "standard time intervals" between two events.

To see how the time of a traveler lapse wrt a static observer, let us now load the light clock on a car with Alice.

---

① The emitter and detector are put extremely close together that we can consider them to be at the same point (in the figure they are drawn separately just for illustration).

Load the light clock into Alice's car

To see how the time of a traveler lapse wrt a static observer,[①] let us now load the light clock in a car with Alice (Fig. 1.8). Using ($\mathbb{R}$), Alice finds that the interval of the light clock is $\Delta t_A = 2D/c$.

Now we have to find out Δt_B and compare it with Δt_A.

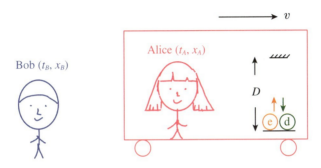

Fig. 1.8

> **Same perpendicular lengths**

You may have an excellent question at this moment: when the light clock is moving, how do we know that the length of the light clock is still D wrt Bob?

Consider a thought experiment: train on rail. When the train is at rest, its wheels has the same interval as the rail width. Now the train moves fast. Does its wheels have larger or smaller width compared to the rail? Neither. Otherwise there would be different type of accidents happening (inconsistent events) that contradicts ($\mathbb{E}$).

Alice's time interval of the light clock measured by Bob

Fig. 1.9 is Bob's view on the moving clock. Here x_B and t_B are the space and time wrt Bob, respectively.

We take three snapshots of time: the emission of the signal at $t_B = t_\text{d}^B$, the time that light is reflected by the mirror $t_B = t_\text{r}^B$, and the time that the light is received by the receiver $t_B = t_\text{d}^B$.

[①] Here we assume that the light clock is close enough to Alice, such that at Alice's reference, the space coordinates of both herself and the clock are $x_A = 0$.

The position and state of the light clock is plotted on the diagram.

In Fig. 1.10, we plot the light trajectory in the moving light clock wrt Bob. Bob must agree that the light has hit the mirror (𝔼). Thus the direction of light must be no longer vertical. And from (ℂ), the speed of light from the emitter to the mirror is still c.

From the Pythagorean theorem, we have

$$\left(\frac{1}{2}c\Delta t_B\right)^2 = \left(\frac{1}{2}v\Delta t_B\right)^2 + D^2 \qquad (1.11)$$

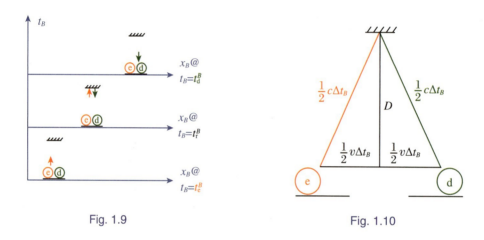

Fig. 1.9 Fig. 1.10

We can thus solve Δt_B — Alice's time interval measured by Bob:

$$\Delta t_B = \frac{2D}{c}\frac{1}{\sqrt{1-\frac{v^2}{c^2}}} = \gamma \Delta t_A, \qquad \gamma \equiv \frac{1}{\sqrt{1-\frac{v^2}{c^2}}} > 1 \qquad (1.12)$$

Thus in Bob's frame, Alice's light clock slows down. This is also known as the "time dilation" of moving clocks.

> **Meaning of Δt_B**
>
> Be reminded that Δt_B is NOT the time duration of Bob's light clock. Rather, it's the time duration of Alice's light clock (i.e. moving with Alice) in Bob's frame. So wrt Bob (static observer), Alice's time duration slows down.

Not only Alice's light clock—but the whole Alice's frame slows down wrt Bob

At this moment, it is not convincing enough to conclude that everything about Alice slows down wrt Bob. Because we have only shown that her light clock slows down[①]. What about her other clocks, her phone, her heart rate ... ?

In fact, according to ($\mathbb{R}$), everything above slows down.

For example, Alice has a phone which can define the unit time interval. She puts the light clock and phone very close to each other. If the phone and the light clock have the same Δt in a static frame, they must have the same Δt_A wrt Alice ($\mathbb{R}$). Their agreement is an event, which Bob must agree ($\mathbb{E}$). Thus Bob must agree that the moving phone defines the same Δt_B as the light clock.

Goodbye to absolute time

By "moving clock slows down", one notes that the concept of "absolute time" by Newton is gone with the speed. Different movers have different times; there is no absolute mover and thus there is no absolute time.

We will come back to the fundamental understanding of time later this part, and later again in general relativity.

Summary and the 4-step reasoning in special relativity

I hope now you have understood why the space traveler Alice is younger than Bob when she returns — as a moving observer, she slowed down wrt Bob.

The 4 key steps of reasoning is important, and we will use similar methods repeatedly. Thus let us summarize it here:

(1) Construct an apparatus in a static frame. The apparatus should be as simple as possible so we can calculate what's actually happening inside the apparatus. Here the apparatus is the light clock with standard time interval $\Delta t = 2D/c$.

(2) Load the apparatus into Alice's moving car. From ($\mathbb{R}$), Alice must find the apparatus work the same way as it is at rest. Here Alice finds $\Delta t_A = \Delta t$.

(3) Calculate what Bob gets. Compare the result with what Alice gets. We have calculated that $\Delta t_B = \gamma \Delta t_A$. Thus the moving light clock slows down.

(4) Although the result is obtained by one particular apparatus, the comparison be-

---

① "I don't care about your light clock. I care about you." — Bob

tween Bob's frame and Alice's frame applies for all possible apparatuses. This is because otherwise, one can use the difference to identify an absolute mover (ℝ).

Frame time and what Bob actually sees

We have mentioned the concept "frame time": if not otherwise stated, when we mention time, we mean time in one's frame, instead of the time that light comes to the observer's eye. Now let us make this point clearer.

Let us put two things at each point in space (i.e. the x-direction, Fig. 1.11):

(1) A space label. For example, imagine there is a ruler along the x-direction and the label of the ruler is the space label.

(2) A clock. The clocks are synchronized (see the section about synchronization for how to synchonize clocks) at all points in space.

Time dilation (Δt_B) means that, when Alice's clock (the clock marked with a v) is moving, at a next moment, it is compared with another clock in Bob's frame, and it slowed down in this comparison. The time dilation does not depend on if Alice's car moves towards or away from Bob.

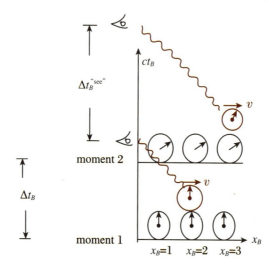

Fig. 1.11

On the other hand, what Bob actually "sees", denoted by $\Delta t_B^{\text{"see"}}$: since the speed of light is finite, there will be a delay for the light ray to travel from Alice's clock to Bob's eyes (the wavy lines). $\Delta t_B^{\text{"see"}}$ depends on whether Alice is moving towards or away from Bob.

Now that you have hopefully understood time dilation and why the space traveler

Alice is younger. So now it is a good time to confuse you again, with the infamous "twin paradox".

Who is moving? Who is younger? — The twin paradox

We have shown: Bob find that Alice is younger (according to the clocks in his frame). However, motion is relative ($\mathbb{R}$). Thus shouldn't Alice also find Bob younger?

Let us answer the question in two different setups:

(1) Alice and Bob are both inertial frames, i.e. they move wrt each other forever. Then from ($\mathbb{R}$), both of the statements are correct: "Alice finds Bob younger" and "Bob finds Alice younger".

At first sight, this contradicts with ($\mathbb{E}$). But in fact there is no contradiction. As inertial movers, Alice and Bob can meet only once. This is the time that Alice starts off with the same age as Bob. Afterwards they never meet again. Thus there is no local events to compare their ages (to compare Δt_B and Δt_A using *events*, they have to meet at least twice).

(2) Alice first moves and later returns to meet Bob. This is the same setup as we discussed at the very beginning of this part.

Then we can only trust Bob at the moment. This is because for Alice to return, there exist a period of time when Alice is not in an inertial frame. The principle of relativity (we used a lot of times when deriving the conclusion) does not apply for non-inertial observers. Thus we can only trust Bob's statement that indeed Alice is younger.

Time travel?

The twin paradox is an example of time travel to the future. Take a spaceship trip and back, you will be in the future. If the spacetrip is fast enough, within 1 year round trip, you can see the earth of the next century, or even later. In general relativity you will see more ways to travel to the future such as stay close to a black hole. What about time travel to the past? Later in this part you will find difficulties within the framework of special relativity to prevent time travel to the past. There are also more general problems, such as what if you travel to the past, and prevent your father and mother to meet? What if you travel to the past and meet another version of yourself? And so on. It is probably impossible to travel to the past, though no decisive conclusion can be made at this moment.

1.3 Physical Picture and Physical Intuition

This section is not about any detailed rule of physics. Nevertheless, considering that modern physics is so different from classical physics, it is important to pause and discuss about method of learning. This is because modern physics is very different from classical physics. We choose to put the section here, because before having any real experience in modern physics, the talk is empty.

Why modern physics is different from classical physics (i.e. general physics)?

From time dilation of special relativity, you have already tasted a tinny little bite of modern physics. If you have not learned it already, I am sure that you feel it is counter-intuitive and more confusing than most parts of general physics.

And much more is coming — you will learn more about relativity, gravity, quanta, information, complexity ... There is a whole new world in front. The common features of these subjects are not only that they were discovered in the past about 100 years, but also, they are farther away from our daily experience. This is the difference from classical physics.

Do not be scared by that! This is why we have this section.

Incidentally, there are similar transitions in mathematics, literature and art — they also evolve from classical eras to modern eras. The modernization revolutions in literature, art, mathematics and physics are of course different, but share interesting similarities.

Let us use an exercise to talk about the methods of thinking, and the recommended way to learn modern physics.

An exercise

A "hand grenade" is a kind of bomb that explode a certain time after you trigger it. Let the time interval between trigger and explosion be Δt when the hand grenade is at rest. Now you throw the hand grenade with speed v see Fig. 1.12. Calculate the distance the hand grenade travels before explosion in the framework of special relativity. In the calculation, both the earth gravity and the friction from air can be neglected.

Instead of solving the excise immediately, let us discuss 3 ways to "think about" how to solve it (*not* three ways to solve it).

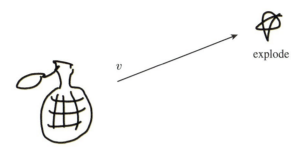

Fig. 1.12

(1) *Pattern matching*. Match (transform) the problem to a solved one: "Bob" ↔ "you"; "Alice" ↔ "grenade"; $\Delta t_A \leftrightarrow \Delta t$ and thus $v\Delta t_B = \gamma v \Delta t \to s$.

(2) *Inverse search*. The distance is asked. How to calculate distance, the speed is given, so $s = vt$. Now how to get t? Which equation has t? It is a problem about relativity so should be $\Delta t_B = \gamma \Delta t_A$ (bless you get γ instead of $1/\gamma$, or use a bit pattern matching). Thus, $s = \gamma v \Delta t$.

(3) *Physical picture guided*. Imagine the grenade is flying a distance and then explode. You see the words but a movie is played in your mind. The time of the grenade is slowing down. So it must be running a longer time than usual, and thus a longer distance. The flight time is longer by a γ factor, and thus the distance travelled is longer by a γ factor: $s = \gamma v \Delta t$.

When preparing exams, we may have trained a lot using mainly the 1st and 2nd methods. However, the 3rd way of thinking is the major direction to go if you'd like to become an innovative physicist, or at least think as a physicist. Because:

Why physical intuition and physical picture are important

(1) A physical picture connects what you learn to the real world. The other two methods do not. In research you need to know the possible applications of the theory, or how to do approximations to simplify the analysis. The physical picture guides you to do that. The other two do not. Some comments from Feynman may be nice to read at this point.

(2) A physical picture tells you if your answer makes sense. If you made a mistake somewhere, you have a good chance to catch it early if you have a physical picture in your mind.

(3) Research means new problems that pattern matching does not give you good results. (But you indeed have to use pattern matching to make sure your idea is new.)

(4) Researchers has to define problems before solving them (unlike exam problems which are already sharply defined). A clear physical picture tells you how to define the problem. Inverse search or pattern matching cannot.

(5) As a researcher, one needs to talk to people. In most discussions (especially those without a blackboard around), you talk by physical pictures.

(6) Computers are good at pattern matching and inverse search. The rise of Artificial Intelligence (AI) is more likely to reduce the value of your pattern matching and inverse search works, but less likely the ones based on intuition.

I hope you are convinced now: Even physics becomes less intuitive in the modern era, you should try to make it as intuitive as possible by building up physical pictures. And here are some suggestions on how to achieve that.

> **Keep a balance**
>
> Having said the importance of intuition, surely intuition is not the only important thing. A working physicist must be able to do professional calculations (though relatively easier to train). Thus you should keep a balance between intuition and calculation. It is important but hard to know whether intuition or calculation is more important for you now — it's always difficult to know oneself. You may ask some good physicists for advices.

How to build up physical picture/intuition in modern physics?

(1) Try to "play a movie" in your mind about the problem. Include as many relevant physical details in the movie as possible. When you have new understandings on something, add it to the "movie".

(2) When you find something not intuitive, think about it again and again. You will eventually feel happier with it.

(3) Pay attention to paradoxes in physics, and how they get resolved.

(4) Think about the same problem in different ways/angles.

(5) Compare with our everyday life experience. Find similarities and/or key differences.

(6) Simplify and modularize the problem. Build up the intuition of the simplest problems as building blocks of more complicated ones.

(7) After you have learned some advanced courses or do research, using a "higher level" view to try building better intuition of more elementary courses that you understand

better.

(8) Repeat known calculations in your mind. Indeed the thing that you repeated is still calculation, but without pen and paper, you will have to simplify the problem, pay less attention to factors of two, and pay more attention to the physical essense.

(9) Carefully verify your intuition with the standard formulae in the textbook. This is very important because your intuition may be wrong and needs to be rebuilt.

(10) If there are still parts which you cannot make it intuitive, use some (as little as possible) math derivation for the moment. Try to replace it by real intuition later.

1.4 Length Contraction

We have just witnessed the revolution of the concept of time — time duration is relative depending on the motion of observers. Now is the turn of space. Is space interval still absolute, or it is similarly relative depending on observers[①]?

In this section, we will arrive at the same conclusion of length contraction in 2 ways. You are expected to understand the first well. The other is optional but recommended as to understanding the same problem in different aspects.

The grenade revisited

Let's think about the grenade problem from a different angle. Recall wrt the ground, the grenade has flown a distance $s = \gamma v \Delta t$. Now, what the grenade "thinks"?

It still has lifetime Δt (wrt itself). It would think that it has flown a distance $v\Delta t = s/\gamma$. But the observer standing on the ground thinks that it has flown a longer distance (wrt a static observer). How can it be?

To make the situation clearer, let us put a ruler on the ground, and let the grenade slide over the ruler (Fig. 1.13). The rest length of the ruler is $s = \gamma v \Delta t$.

[①] In fact, length contraction was hypothesized way before special relativity, by FitzGerald (1889) and Lorentz (1892) to explain the Michelson-Morley experiment. For this reason it is also known as Lorentz contraction. But it is special relativity that put length contraction into a solid and general context.

Wrt the grenade, the ruler (and ground) is moving with v for time Δt. Thus the ruler has length s/γ wrt the grenade.

Grenade sliding on a ruler:
Triggered at one end and exploded on the other.
Rest length of ruler: $S=\gamma v \Delta t$
Wrt grenade: Length of ruler is $v\Delta t = S/\gamma$.
Fig. 1.13

To conclude, a moving ruler is shorter by $1/\gamma$ if placed parallel to the motion direction. Also recall that if the ruler is perpendicular to the motion direction, then the length does not change.

We would have closed up the section here. But what happened during Δt time in a grenade is opaque. Let us construct a light ruler to see what happens explicitly, which can also give insight on velocity addition.

(Optional) Length contraction from the light ruler (4-step reasoning)

Although we have got the result, it is a good practice to derive it using another way. Let us construct a light ruler and study how it contracts. This will explicitly verify our promise: No matter how the ruler is defined, it should contract the same way. To see that, let us apply our familiar 4 steps:

(1) In a static frame, the length of the ruler is $D = \frac{1}{2}c\Delta t$, where Δt is the time interval between the emission and detection of light.

(2) Now we load the light ruler on Alice's car. The situation and the spacetime diagram that Bob finds are in Fig. 1.14.
According to (ℝ), Alice consider the length of the ruler to be

$$D_A = \frac{1}{2}c\Delta t_A \tag{1.13}$$

(3) What length shall Bob get? We first divide the time interval Δt_B into two parts: Δt_{B1} and Δt_{B2} for the time interval from the emitter to the mirror and from the mirror to the detector, respectively. From the above figure (right panel),

$$c\Delta t_{B1} = D_B + v\Delta t_{B1}, \qquad c\Delta t_{B2} = D_B - v\Delta t_{B2} \tag{1.14}$$

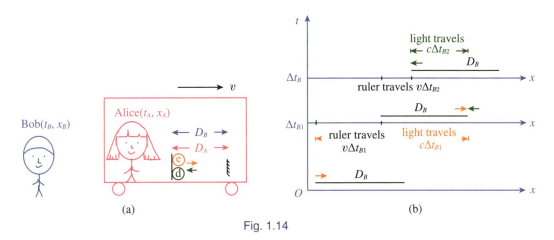

Fig. 1.14

Thus

$$\Delta t_B \equiv \Delta t_{B1} + \Delta t_{B2} = \frac{D_B}{c-v} + \frac{D_B}{c+v} = \frac{2}{c}\frac{D_B}{1-\frac{v^2}{c^2}} = \frac{2}{c}\gamma^2 D_B \quad (1.15)$$

where as always, $\gamma \equiv \frac{1}{\sqrt{1-v^2/c^2}}$. Thus

$$D_B = \frac{c}{2}\frac{\Delta t_B}{\gamma^2} = \frac{c}{2}\frac{\gamma \Delta t_A}{\gamma^2} = \frac{c}{2}\frac{\Delta t_A}{\gamma} = \frac{D_A}{\gamma} \quad (1.16)$$

Again, we arrive at the conclusion that moving ruler contracts in the motion direction.

(4) For all rulers, one should get the same conclusion as the light ruler.

Meaning of D_B

Be reminded that D_B is NOT the length of Bob's ruler. But rather, it's the length of Alice's ruler (i.e. moving with Alice) in Bob's frame. So wrt Bob (static observer), Alice's ruler (moving ruler) contracts.

(Optional) A first look at velocity addition

Let's slightly modify the light ruler: replace the light from emitter to mirror into a moving particle, with speed u_A wrt Alice. After the "mirror" gets the particle, the "mirror" still sends back a beam of light (thus it shouldn't actually be called a mirror though). Then how the light ruler experiment get modified?

(1) Wrt Alice, for the particle forward, $u_A \Delta t_{A1} = D_A$; and for light moving back,

$c\Delta t_{A2} = D_A$. Thus,

$$\Delta t_A = \Delta t_{A1} + \Delta t_{A2} = D_A \left(\frac{1}{u_A} + \frac{1}{c} \right) \tag{1.17}$$

A speedmeter

What's the motivation to replace c with v in one way? We can ask what this apparatus can do when it's at rest. From $\Delta t = D/u + D/c$, as we have defined Δt and D, we can solve u. Thus, the apparatus is a speedmeter. No wonders that a speedmeter can tell you about speed addition.

(2) Wrt Bob, he will find the particle moving at a different velocity u_B. For the particle moving forward, $u_B \Delta t_{B1} = D_B + v \Delta t_{B1}$; and for light moving back, $c\Delta t_{B2} = D_B - v\Delta t_{B2}$. Thus,

$$\Delta t_B = \Delta t_{B1} + \Delta t_{B2} = D_B \left(\frac{1}{u_B - v} + \frac{1}{c+v} \right) \tag{1.18}$$

(3) We already know that $\Delta t_B = \gamma \Delta t_A$ and $D_B = D_A/\gamma$. Now divide Eq. (1.18) by Eq. (1.17) (LHS and RHS, respectively), we get

$$u_B = \frac{u_A + v}{1 + u_A v/c^2} \tag{1.19}$$

When taking $u_A \to c$ limit, we find that $u_B \to c$, consistent with (ℂ). In section 1.7, you will learn a more general version of the velocity addition formula.

The additive rapidity

Eq. (1.19) looks ugly. Wouldn't nature be simpler? But who told us velocity is the best variable to describe motion? Let us define *rapidity*: $\phi(v) \equiv \text{arctanh}(v/c)$. Inserting this definition to Eq. (1.19), we simply get

$$\phi(u_B) = \phi(u_A) + \phi(v)$$

Thus rapidity is the variable that is actually additive.

Why hyperbolic functions arises here (they originally are functions to parameterize a hyperbolic curve $x^2 - y^2 = 1$ with additive parameters, just as what trigonometric functions sin, cos, tan are functions to parameterize a circle $x^2 + y^2 = 1$ with additive angles)? Do the hyperbolic functions imply a new underlying math structure? Why motion in such a formula looks not like division of space and time, but rather hyperbolic rotation of space and time? We will return to this question in section 1.8.

We have got two results from the light clock now: time dilation; and after time dilation have become known, we rotate the light clock to get length contraction. Now that length contraction have become known, can we get something else?

1.5 Meaning of the "Same Time" (Simultaneity)

Moving ruler contracts. Now, is there a twin ruler paradox? If Alice and Bob each holds a ruler along their motion direction, and when they meet, they compare the length of their ruler, can they find each other's ruler shorter? How can that happen?

The key observation is that, to be fair, they have to compare the two ends of the ruler at the same time. Wait, do they have the same concept of the same time?

1.5.1 Simultaneity Depends on Which Observer

The meaning of simultaneity wrt an observer

Consider two small objects P and Q, not moving wrt each other (Fig. 1.15). An observer is standing exactly at the midpoint of PQ, and not moving wrt PQ. This can be practically done with a static ruler: Let P be at the 0 m point, Q be at the 1 m point, and the observer at the 0.5 m point.

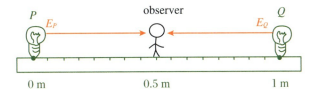

Fig. 1.15

> **Why introducing two objects?**
>
> Actual objects P, Q are needed to find an observer standing at the midpoint. Because a midpoint is defined for two points in space (P and Q), not for two events (E_P and E_Q) in spacetime.

Introduce two events: event E_P happens to object P; and event E_Q happens to object Q. For example, E_P and E_Q are the turn-on time of light bulbs at P and Q, respectively. Recall that each event happens at a particular moment.

Now we are ready to define whether E_P and E_Q happens at the same time, or one is earlier/later than the other, wrt the observer:

If the light signal from E_P and E_Q reach the observer at the same time, then E_P and E_Q happens at the same time. Otherwise, whichever reaches the observer earlier happens earlier.

> **Relating to the time of a frame**
>
> Previously, we have introduced the time of a reference frame — at different positions, time are synchronized using light signals, deducting the time used for light propagation. Here, the concept of simultaneity means that E_P and E_Q happens at the same coordinate time in this static frame.

1.5.2 A Spacetime Diagram of the Above Scenario

Let us draw the events in a "spacetime diagram" (Fig. 1.16). Spacetime diagrams will turn out to be useful tools in studying relativity. On a spacetime diagram:

(1) An event is a point.

(2) Light travel 45° lines.

(3) An object (or observer, except light) is a line (called world line) with |slope| > 45° everywhere.

(4) An inertial observer is a straight line.

(5) A static object is a line parallel to the ct axis.

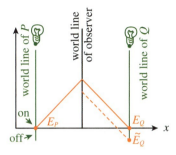

Fig. 1.16

(6) Events at the same time are parallel to the x axis.

> **Am I wasting your time?**
>
> Oh, it looks that I am wasting your time by explaining something you already know since kindergarten. This is true. However, one little step further will need university education — load the device into a car.
>
> Einstein said with great modesty: "How it happened that I in particular discovered the relativity theory, it seemed to lie in the following circumstance. The normal adult never bothers his head about space-time problems. Everything there is to be thought about it, in his opinion, has already been done in early childhood. I, on the contrary, developed so slowly that I only began to wonder about space and time when I was already grown up. In consequence I probed deeper into the problem than an ordinary child would have done."

If the light rays connecting E_P and E_Q reaches the world line of the observer at the same point, they happen at the same time wrt the observer. On the contrary, $\widetilde{E}_Q$ is considered earlier than E_P wrt the observer as the light ray from $\widetilde{E}_Q$ arrives earlier at the observer.

Simultaneity is a relative concept (4-step reasoning)

(1) We have defined simultaneity for a static observer with the above P-observer-Q system.

(2) Let us load the system into a car and let the moving observer be Alice. Let us study the events E_P and E_Q, which happen at the same time wrt Alice. In our example, that corresponds to the light bulbs turning on at the same time at P and Q wrt Alice. See left panel of Fig. 1.17.

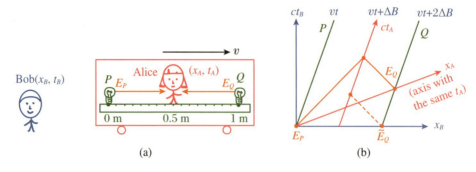

Fig. 1.17

(3) Wrt Bob, do E_P and E_Q happen at the same time? In other words, do E_P and E_Q have the same time coordinates in Bob's frame? Making use of a spacetime diagram in Bob's frame (right panel of Fig. 1.17), we immediately find that E_P is earlier than E_Q wrt Bob. On the contrary, an event $\tilde{E}_Q$ considered to be at the same time with E_P wrt Bob, is considered earlier than E_P wrt Alice.

(4) The relativity of simultaneity is not only true for the P-observer-Q system, but for all consistent definitions of simultaneity ($\mathbb{R}$).

> **Steps to draw Bob's spacetime diagram**
>
> Draw Bob's axes and the world lines of P, Q, and Alice's position (i.e., the ct$_A$ axis).
> Draw the event that two beams of light meet Alice.
> Draw the history of these two beams of light as 45° lines.
> The intersection of light and P, Q are events E_P, E_Q.
> Compare the time of E_P and E_Q in Bob's frame.
> Draw the spacetime axes of Alice's frame in Bob's frame.

Now you should be able to resolve the puzzle of "who wrote the letter first".

Recap: equal time slices

To condense the above analysis into one figure (Fig. 1.18): When Alice is moving wrt Bob, the coordinate system of Alice drawn in Bob's coordinate system is as the figure to the right. It is similar to rotation, but both space and time axes oddly fold inwards. We will discuss this transformation (Lorentz transformation) in more details later. You should pay special attention on the equal time lines wrt Alice on this figure — different from that wrt Bob.

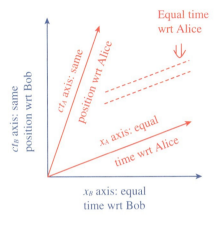

Fig. 1.18

Twin paradox revisited

In our twin paradox, Alice is not an inertial observer at the turn around time. Thus she cannot use special relativity of an inertial observer to *directly* explain her experience. However, Bob can help her to figure out what happens at the turn-around Fig. 1.19:

Before and after the turn-around, Alice is in two *different* inertial frames. Bob's age "jumps" when Alice switches from the before-turn-around frame to the after-turn-around frame. Thus in Alice's frames, there is a sudden change in Bob's age.

As mentioned, to describe what Alice sees (light arrives in her eyes), light propagation time needs to be added. I will leave the details for your exercise.

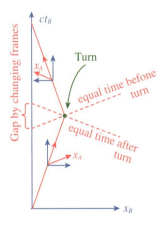

Fig. 1.19

Can SR describe acceleration?

There is a common misconception saying "special relativity cannot describe the experience of an accelerating observer". This is *wrong*. The laws of special relativity (time dilation, length contraction, and more later) are expressed in inertial frames (just because the laws are mathematically simpler in inertial frames). Thus, we need to *use an inertial frame* to apply these laws to our question. However, we can *calculate in this inertial frame* what an accelerating observer sees (how light reach her eyes, for example). In this way, the experience of an accelerating observer is described *with the help of an auxiliary inertial frame*.

1.5.3 Causality and Types of Separations

Now we have understood: the concept of simultaneity is relative to observers. For example, Alice and Bob may consider differently on who wrote the letter first. In other words, time orders of some events (here two events: Alice writes her letter; Bob writes his letter) are reversible.

A natural question then is: Are all time orders between events reversible?

Time order associated with cause and effect

Fig. 1.20

Consider an example: Lightning strikes on a tree (Fig. 1.20), and then the tree dies. There are two events here:

(1) Event E(strike): Lightning strikes on the tree.

(2) Event E(die): The tree dies. This is an *effect*, *caused* by E(strike).

May a moving observer Ms. Bright observe that E(die) happens earlier than E(strike)?

The cause-effect relation (known as causality) is at the heart of physics. Physics is about prediction of how an initial state evolves with time. In other words, the cause-effect relation is how questions get explained in physics — "Why (effect)? Because (cause)."

> **Correlation**
>
> Cause-effect is a special case of correlation (conditional probability). If events E1 and E2 have nontrivial correlation, maybe
>
> (1) E1 causes E2.
>
> (2) E2 causes E1.
>
> (3) Both E1 and E2 are caused by something else.

As causality is so important, we hope that it is preserved in special relativity. Happily, special relativity is indeed *consistent* with causality. Having that said, causality is an independent postulate added to special relativity. In other words, special relativity itself does not *derive* causality. Let us now see how causality and relativity work together.

Special relativity without superluminal motion is consistent with causality

Spacetime diagram of subluminal Ms. Bright and superluminal Ms. Bright (Fig. 1.21). For any subluminal observer, the world line of the tree has slope greater than 45°. Then, as long as Ms. Bright moves no faster than the speed of light, she cannot flip the order of E(strike) and E(die).

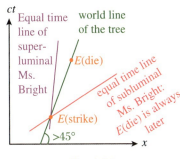

Fig. 1.21

No superluminal lightning either

For causal order to be respected, not only Ms. Bright is not supposed to travel faster-than-light; but also no lightning (or any information causing the death of the tree) can be superluminal either. Otherwise causality is violated even with subluminal Ms. Bright.

Buller's poem (1923)

There was a young lady named Bright;

Whose speed was far faster than light;

She set out one day; In a relative way;

And returned on the previous night.

We hope this to be forbidden. Otherwise, what if Ms. Bright returned to the previous night, and locked herself in the room, how can her superluminal trip happen at all?

So far, the most convincing explanation is that, superluminal trip should be forbidden. Having that said, the possibility of a time machine is an open question for modern physics.

Thus, if superluminal movers are forbidden, then the cause-effect order is preserved. In this course, we will assume that superluminal motion is indeed forbidden in special relativity. This is consistent with all known experiments.

More generally, no information can be sent faster than the speed of light. Because information must be sent by matter after all. And from the consistency of the theory, information can bring causal connection between events. Thus if information can be sent faster than light, then the same problems of superluminal motion can arise.

You may have an excellent question at this point: What if we push Ms. Bright so she accelerates from subluminal to superluminal? It is in fact impossible. We will come back to this point later.

In general relativity, you may hear that things can go superluminal, for example, for cosmic expansion. This is in some sense right while in some other sense wrong. One has to first define velocity precisely. We will come back this point at the end of this course, in the cosmology section.

Causal structure of spacetime: past and future light cones

From separation of events, given a spacetime point (say, you at the present time), the spacetime is divided into three different regions according to the causal connection to the observer (Fig. 1.22).

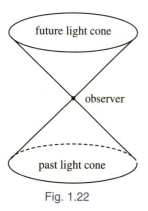

Fig. 1.22

(1) Past light cone. This region contains all the causes of you up to now. You have not yet seen anything beyond this region.

(2) Future light cone. This region contains all the effects by you from now on. You can no longer change anything beyond this region.

(3) Outside both past and future light cones: there is no cause-effect relations between the present version of you and any point there.

No perfectly rigid body in relativity

Let's consider a thought experiment: Alice and Bob are separated by 5 light years. And they hold a 5-light-year-long rod, which is a perfect rigid body (Fig. 1.23). Then can Alice and Bob send information faster than light by pushing the rod?

The answer is a simple straight no. Because no information can be faster than light,

Fig. 1.23

perfectly rigid body does not exist in special relativity.

If this answer is too brute, we can also see dynamically what happens. The rod is (usually) made of atoms and the force propagating between atoms need at least speed of light to react a push.

The speed of light limit classifies the intervals between two events into 3 classes.

Space-like, null and time-like intervals

There are three types of intervals between events (Fig. 1.24).

Space-like: with slope $< 45°$. There exists observer wrt whom two space-like separated events happen at the same time. The events are thus pure space separated wrt this observer. The time order of the events can be flipped for different observers.

Time-like: with slope $> 45°$. There exists observer wrt whom two time-like events happen at the same position. The events are thus separated only in time wrt this observer. The time order of the events is absolute and has to be agreed on for all observers.

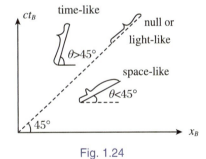

Fig. 1.24

Light-like (or called null): the boundary separating space-like and time-like intervals, with slope $45°$. Light travels with light-like lines.

1.6 Example: The Ladder Paradox

Let's apply what we have learned to get confused (or hopefully not).

The ladder and the garage

As shown in Fig. 1.25, Consider a 15 m long ladder (the ends are labeled P and Q). Alice holds it and moves it towards a garage, with speed 0.87c, i.e. $\gamma = 2$. The garage is 10m long, and the front door and back door are labeled F and B, respectively. Bob sits static with the garage.

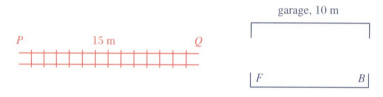

Fig. 1.25

What does Alice find? Alice finds that the garage moves, and thus the length of the garage is $10/2 \text{m} = 5$ m (Fig. 1.26). Thus the ladder does not fit in the garage.

What does Bob find? Bob finds that the ladder moves, and thus the length of the ladder is $15/2 \text{m} = 7.5$ m (Fig. 1.27). Thus the ladder fits in the garage.

Fig. 1.26   Fig. 1.27

Who is right?

Both are right.

Note that "ladder fit in garage" is not an event which localized at a spacetime point. Especially, there are two important events (Fig. 1.28):

① P enters front door F.

② Q exits back door B.

If ① is earlier than ②, then the ladder fits in the garage. If ② is earlier than ①, then the ladder does not fit in the garage.

Wrt Alice, ② is earlier than ① (does not fit); Wrt Bob, ① is earlier than ② (fits).

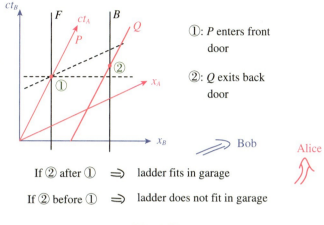

Fig. 1.28

The ladder and the garage with doors

Let's add a local event to sharpen what happens (Fig. 1.29). Let the back door of the garage always be closed. Bob closes the front door once he finds the ladder moves in the garage.

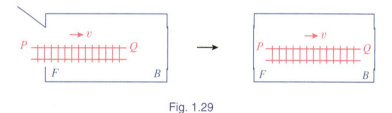

Fig. 1.29

Now even Alice[①] has to agree that the ladder is completely in the garage because the doors are closed. So the "paradox" becomes sharper: Alice thinks that the ladder is longer than the garage; but it is now completely in the garage!

Resolution: Remember that in relativity, there is no perfectly rigid body.

Assuming that the door is relatively stronger and does not break, then the ladder must fall into parts. Wrt Alice: when Q stopped moving, the "stop" information has to be passed to P and it takes at least time 15 m/c. During this time, the front door moved about 13 m. Thus P is well inside F by this time. Of course, it fits into the 5 m garage very well.

On the other hand, if the ladder is relatively stronger and does not break, then the back door is broken and there is no difference from no back door.

[①] Here we assume that Alice is still in an internal frame, no matter any part of the ladder may stop or not.

Replacing the garage by a trap

Let us further replace the binary doors by boundary of a trap (Fig. 1.30).

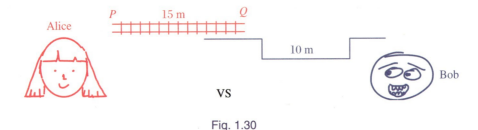

Fig. 1.30

Bob considers that Alice with the ladder start to fall in when the ladder is entirely in the trap. Thus Alice falls in. However, Alice considers herself still not yet fall-in before the Q point escapes. Who's right?

This really depends on how Alice holds the ladder. If Alice holds the ladder perfectly parallel to the ground and sliding on the ground, then Alice falls in. Because there is no perfect rigid body. When Q falls into the trap, Q inevitably falls downwards. However, if Alice holds the ladder in the way a bit upwards, such that Q does not need ground support from the beginning, then Alice can escape.

Replacing the garage by a circuit

Let us stop destroying things and go electronic. Consider the situation below (Fig. 1.31).

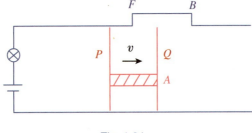

Fig. 1.31

Alice holds two moving conducting wires (they are isolated wrt each other). Both can close the circuit. Wrt Bob, there exists a time when neither of Alice's wires closes the circuit so the bulb shuts off for a moment. Wrt Alice, at least one wire touches the circuit all the time. So shall the bulb shut off?

The key to resolve the problem is that, electricity needs time to conduct. The speed of electricity is comparable to c, but still less than c. Thus wrt Alice, to keep the bulb on without a off-time gap, the electric current has to reach F before Q passes F. So there is an off-time wrt Alice, too.

1.7 The Lorentz Transformation

Alice and Bob need a rule for their coordinate transformations

An event P happened. Both the moving Alice and the standing Bob agree on the existence of the event. However, the same event has different coordinates in Alice's and Bob's frames.

Given the coordinate of P: (t_B, x_B) in Bob's frame, how to get the coordinate (t_A, x_A) for the same event in Alice's frame (Fig. 1.32)?

Recall that the transformation between inertial frames is called a boost. Thus the question can also be asked as: how does coordinates transform under a boost?

Before solving this problem, let us draw similarity between boost and rotation (Fig. 1.33).

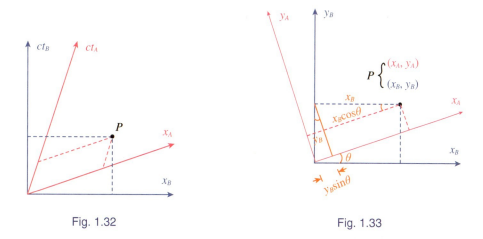

Fig. 1.32　　　　　　　　　　Fig. 1.33

The case of rotation: Alice and Bob are not moving wrt each other, but facing different directions. Given Bob's coordinate of an object (x_B, y_B), how to get Alice's coordinate for the same object (x_A, y_A)? Following relations of Euclidean geometry, we get:

$$\begin{cases} x_A = x_B \cos\theta + y_B \sin\theta \\ y_A = -x_B \sin\theta + y_B \cos\theta \end{cases} \quad (1.20)$$

I hope you get a better idea about the question by this analogue. Also, in the next section we will find a surprising similarity between these space-time and space-space transformations.

> **Velocity in a general direction**
>
> Be reminded that the relative velocity v between frames is only along the x direction (and positive for Bob). If you are dealing with velocities in a general direction $\boldsymbol{v} = (v_x, v_y, v_z)$, you can first do a rotation to rotate it to the x direction before applying the transformation.

Space transformation of a boost

Given Bob's coordinate (t_B, x_B), now we would like to solve x_A. How to do that? An equivalent question is that, how to physically describe x_A? Note that x_A is Alice's spatial coordinate. Thus, if Alice carries a ruler (static wrt Alice) with length x_A, Alice holds one end, then the other end has coordinte x_A (Fig. 1.34).

For simplicity, we choose the same origin of coordinate for Bob and Alice (Fig. 1.35). Given Bob's coordinate of P: (t_B, x_B), how can we find the relation $x_A(t_B, x_B)$?

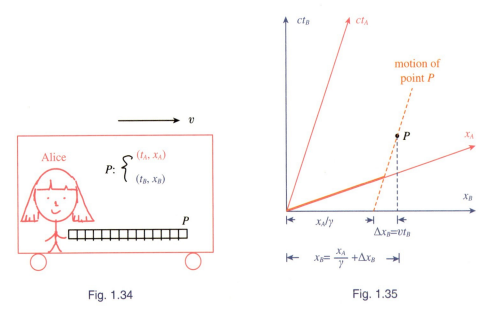

Fig. 1.34   Fig. 1.35

Wrt Bob, the moving ruler has length x_A/γ. Thus at Bob's time $t_B = 0$, the point P has $x_B = x_A/\gamma$. However, at $t_B \neq 0$, the end point P has also moved a distance vt_B. Thus for general t_B, $x_B = x_A/\gamma + vt_B$. In other words, we get $x_A(t_B, x_B)$ (i.e. x_A as a function of t_B and x_B) as

$$x_A = \gamma(x_B - vt_B) \tag{1.21}$$

How to get $x_B(t_A, x_A)$? We can flip Alice's and Bob's role and do the calculation again. But in fact we do not have to do it. Simply following (ℝ), we get

$$x_B = \gamma(x_A + vt_A) \tag{1.22}$$

Time transformation of a boost

The remaining question is: How to get $t_A(t_B, x_B)$ and $t_B(t_A, x_A)$? We do not have to do more thought experiments since we can solve them from Eq. (1.21) and Eq. (1.22).

Inserting Eq. (1.21) into Eq. (1.22) to eliminate x_A,

$$x_B = \gamma[\gamma(x_B - vt_B) + vt_A] \quad \rightarrow \quad t_A = \gamma\left(t_B - \frac{v}{c^2}x_B\right) \tag{1.23}$$

Similarly, or simply using the principle of relativity, we get

$$t_B = \gamma\left(t_A + \frac{v}{c^2}x_A\right) \tag{1.24}$$

Lorentz vs Einstein

The Lorentz transformation is named after Lorentz for his work during 1892—1904. In other words, the Lorentz transformation is known before Einstein's special relativity (1905). It was discovered as mathematical symmetries behind the Maxwell equations and an phenomenological approach to explain the Michelson-Morley experiment. But it was Einstein who first understood the physics of such transformations.

Summary: the Lorentz transformation

Let's summarize the transformation we have got, and add back the trivial y and z directions. These rules to transform between inertial frames is known as the Lorentz transformation. Here $\beta \equiv v/c$.

$$\begin{cases} ct_A = \gamma(ct_B - \beta x_B) \\ x_A = \gamma(x_B - \beta ct_B) \\ y_A = y_B \\ z_A = z_B \end{cases} \qquad \begin{cases} ct_B = \gamma(ct_A + \beta x_A) \\ x_B = \gamma(x_A + \beta ct_A) \\ y_B = y_A \\ z_B = z_A \end{cases} \qquad (1.25)$$

The Lorentz symmetry

Under the appearance of a transformation, a more physical name to refer to Eq. (1.25) is the Lorentz symmetry: Alice and Bob are symmetric, in that though they use different coordinates to describe events, they observe the same laws of nature (for example, equations of motion for particles). Symmetry is now considered the first principle in fundamental physics. We will come back to the concept of symmetries in the part of action principle.

The Lorentz transformation describes the mathematical structure of special relativity. Thus our known results can be directly read-off from the formulas (1.25).

Time dilation, length contraction and simultaneity revisited

Simultaneity: Consider two events E_1 and E_2, happened at the same time wrt Alice, i.e. $t_A^{E_1} = t_A^{E_2}$. Eq. (1.25) gives $0 = c(t_A^{E_1} - t_A^{E_2}) = \gamma[c(t_B^{E_1} - t_B^{E_2}) - \beta(x_B^{E_1} - x_B^{E_2})]$. Clearly, if the two events happens at different locations and if $\beta \neq 0$, then the two events are not at the same time wrt Bob.

Time dilation: A clock is moving wrt Alice with coordinate $(t_A, 0)$. Note that $x_A = 0$ here. Since Alice carries the clock, the clock is always located at the origin wrt Alice. From the first equation to the right panel of (1.25), we get $t_B = \gamma t_A$.

Length contraction: A ruler is moving together with Alice. How is Bob supposed to measure its length? Bob should measure his coordinates of both ends of the ruler *at the same time*. The left end of the ruler is at $x_B = 0$ when $t_B = 0$. Thus the coordinate of the ruler at $t_B = 0$ is the length of the ruler. From the second equation to the left panel of (1.25), we get $x_B = x_A/\gamma$.

> ### Why don't you tell us earlier?
>
> You may feel furious here:
>
> "I have spent great efforts in understanding time dilation, length contraction and simultaneity. But now they follow so simple equations. Why don't you start from telling us the Lorentz transformation and derive everything from there? That would have saved a lot of my time."
>
> I am in fact pretty careful in choosing the point to introduce Lorentz transformation.
>
> If I introduce it earlier, these effects will be just cold math formulas in your mind instead of living characters (i.e., with a physical picture).
>
> On the other hand, I can teach velocity addition, energy and momentum first and put the Lorentz transformation to the very end. But then the physical scenarios become too complicated to be helpful. I thus choose here to be the point to introduce this powerful tool.
>
>

Now we are ready to solve a problem: What's wrong with the Newtonian velocity addition rule $\boldsymbol{v}_B = \boldsymbol{v} + \boldsymbol{v}_A$ when $v_A \sim c$? In the optional material of the length contraction section, we studied a special case. Here we will derive general formulas.

Addition of velocity

For simplicity, let us take the relative velocity $\boldsymbol{v}$ to be $(v,0,0)$, along the x-direction, as we have always assumed.

Now Alice throw a ball with velocity

$$\boldsymbol{v}_A = (v_{Ax}, v_{Ay}, v_{Az}) = \left(\frac{\mathrm{d}x_A}{\mathrm{d}t_A}, \frac{\mathrm{d}y_A}{\mathrm{d}t_A}, \frac{\mathrm{d}z_A}{\mathrm{d}t_A}\right) \tag{1.26}$$

and Bob finds

$$\boldsymbol{v}_B = (v_{Bx}, v_{By}, v_{Bz}) = \left(\frac{\mathrm{d}x_B}{\mathrm{d}t_B}, \frac{\mathrm{d}y_B}{\mathrm{d}t_B}, \frac{\mathrm{d}z_B}{\mathrm{d}t_B}\right) \tag{1.27}$$

Now use the rule of Lorentz transformation, one can calculate

$$v_{Bx} = \frac{\mathrm{d}x_B}{\mathrm{d}t_B} = \frac{\gamma \mathrm{d}(x_A + vt_A)}{\gamma \mathrm{d}\left(t_A + \frac{x_A v}{c^2}\right)} = \frac{\mathrm{d}x_A + v\mathrm{d}t_A}{\mathrm{d}t_A + \frac{v}{c^2}\mathrm{d}x_A} = \frac{v_{Ax} + v}{1 + \frac{v_{Ax} v}{c^2}} \tag{1.28}$$

$$v_{By} = \frac{\mathrm{d}y_B}{\mathrm{d}t_B} = \frac{\mathrm{d}y_A}{\gamma \mathrm{d}\left(t_A + \frac{v}{c^2}x_A\right)} = \frac{v_{Ay}}{\gamma\left(1 + \frac{v_{Ax} v}{c^2}\right)} \tag{1.29}$$

Similarly,

$$v_{Bz} = \frac{v_{Az}}{\gamma\left(1+\frac{v_{Ax}v}{c^2}\right)} \quad (1.30)$$

Example: velocity addition in perpendicular directions

When $\boldsymbol{v}_A = (0, u, 0)$, perpendicular to the relative motion direction, we get

$$\boldsymbol{v}_B = (v, u/\gamma, 0) \quad (1.31)$$

Note that Alice and Bob measures different velocities even in the perpendicular direction, because their time intervals are different (while the length intervals are the same).

1.8 The Geometry of Spacetime

Pythagoras theorem and modern geometry

Since Gauss and Riemann, mathematicians have realized that different types of geometry could be classified by how the Pythagoras theorem appeared on those geometries. For example, on flat, spherical and hyperbolic surfaces, the Pythagoras theorem appears differently (Fig. 1.36).

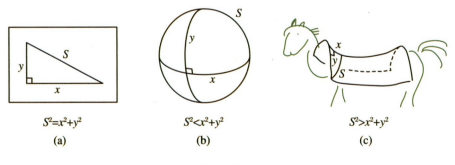

Fig. 1.36

Through more careful studies on how the spherical and hyperbolic surfaces are curved, one can make the above expressions more precise and differentiate the relation twice to define spatial curvature.

Here we are not following this path to study pure spatial geometries in depth. Rather, we would like to ask the following question: now that space and time are "unified" by Lorentz transformation, what does the spacetime geometry look like? Or what does the Pythagoras theorem look like for spacetime?

Our spacetime is NOT Euclidean

Let us reconsider time dilation. Alice carries a clock and both Alice and Bob measure the tick-tock interval (Fig. 1.37).

Wrt Bob, $\Delta t_B = \gamma \Delta t_A$. What does this imply on the spacetime diagram?

Naively, from the geometry, we would have expected $c\Delta t_A > c\Delta t_B$, because $c\Delta t_A$ is the hypotenuse of the right triangle. However, from $\gamma > 1$, we can see $c\Delta t_A < c\Delta t_B$. How is this possible?

As we have discussed, we cannot take the Euclidean geometry for granted.

Especially, in Fig. 1.38, where a and c are time directions (of Alice or Bob), the Pythagorean theorem $a^2 + b^2 = c^2$ no longer holds. Instead, for time dilation,

$$-a^2 + b^2 = -c^2 \tag{1.32}$$

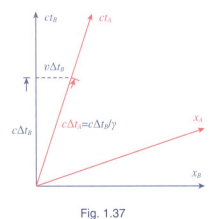

Fig. 1.37

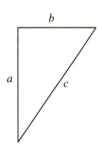

Fig. 1.38

Extra minus signs emerge in front of the square of time, but not space. [1]

---

[1] Note that minus sign in time squared indicates an extra factor of i in front of time to match the Pythagoras theorem of flat space.

Is it a coincidence, or is it in general a new type of geometry?

> **Hyperbolic functions**
>
> Before moving on, let me introduce (or remind you) the definition and features of hyperbolic functions.
> $$\cosh x \equiv \frac{e^x + e^{-x}}{2}$$
> $$\sinh x \equiv \frac{e^x - e^{-x}}{2}$$
> $$\tanh x \equiv \sinh x / \cosh x$$
> $$\cosh^2 x - \sinh^2 x = 1$$
> $$\cosh(i\theta) = \cos \theta$$
> $$\sinh(i\theta) = i \sin \theta$$

Our spacetime is ALMOST Euclidean

To see what has happened in another way, let us rewrite the Lorentz transformation using rapidity $\phi = \text{arctanh}\beta$. Note that $\cosh \phi = \gamma$, and $\sinh \phi = \beta\gamma$. The Lorentz transformation is then

$$\begin{cases} ct_B = ct_A \cosh \phi + x_A \sinh \phi \\ x_B = ct_A \sinh \phi + x_A \cosh \phi \end{cases} \quad (1.33)$$

Do you find it a bit similar to rotation?

Now let's apply a mathematical trick. Define

$$\phi \equiv i\theta, \quad ct \equiv iw \quad (1.34)$$

We then get [1]

$$\begin{cases} iw_B = (iw_A)\cos \theta + x_A(i\sin \theta) \\ x_B = (iw_A)(i\sin \theta) + x_A \cos \theta \end{cases} \Rightarrow \begin{cases} w_B = w_A \cos \theta + x_A \sin \theta \\ x_B = -w_A \sin \theta + x_A \cos \theta \end{cases} \quad (1.35)$$

This is *exactly* a rotation of the w-x plane, with w considered as an extra spatial dimension!

To summarize: A mathematical trick of redefining variables makes Lorentz transformation identical to rotation.

---

[1] Do not ask me the "physical meaning" of the imaginarized time w at the moment. Just consider it as a trick in math. In fact, there are profound implications of imaginary time in quantum field theory. But we are far not ready to introduce them here.

The symmetry of spacetime with imaginary time

According to Pythagoras (570BC—495BC), sphere is the most beautiful and perfect object — it is symmetric under 3 dimensional rotations.

In this respect, our spacetime is even more beautiful than the most beautiful object.[①] This is because, now w, x, y, z directions are no different from each other — they are all related by rotation. Our spacetime is symmetric (i.e., laws of nature have the same form) under rotations in planes w-x, w-y, w-z, x-y, x-z, y-z, and all kinds of combinations of them. It is indeed more symmetric than a sphere because a sphere is only invariant under rotations in x-y, x-z, y-z planes and their combinations.

This further confirms that the difference between space and time is imaginary — once you use imaginary time, time becomes no different from just another spatial direction, and the 3 boosts along x, y and z directions just become three additional rotations.

Natural units — the units of nature

In physics, some units are more natural than the others, because they use the same unit for the same things (or apparently different things, but with the same physical origin).

Let's consider an example of a not-so-natural unit. In ancient China, sometimes people use a different unit for height of mountains (Fig. 1.39). At that time, height seems different from other directions because of gravity. But now we see that the difference is due to the presence of the earth, which spontaneously break the symmetry of empty space. Fundamentally, height has no difference from the other two directions, and they can be related by rotation. Thus, we had better to use the same unit to measure height and the rest of the spatial dimensions.

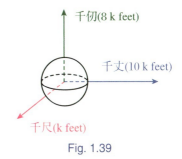

Fig. 1.39

Now that we have seen space and time are not that different and can be related by Lorentz transformation, why we still give space and time different units? We don't have to.

To use the same unit for space and time, we can just set $c = 1$ and then speed is dimensionless.

This is known as the natural unit (together with $\hbar = k_B = 1$). Natural unit is widely used in theoretical high energy physics.

[①] We are not saying that spacetime is a sphere. Rather sphere is curved and our spacetime is flat in special relativity. However, its symmetry includes that of a sphere, and more (there are 4 additional spacetime translations).

Having that said, we shall not use natural units in this class and will always keep c explicit.

> **Natural vs convenient**
>
> The laws of nature are shorter in natural unit as unnecessary conversion of units are avoided. However, whether it is a convenient choice of unit depends on the physical context that we study. If we study explicit dynamics of systems with $v \ll c$, then we had better to keep c explicit because otherwise we will deal with large/small numbers everywhere. For example, for $v = 10^{-8}c \approx 3$ m/s, in natural unit the object only moves 10^{-8} unit of space in one unit of time.

The invariant interval

In Euclidean space,

$$\mathrm{d}s^2 = \mathrm{d}w^2 + \mathrm{d}x^2 + \mathrm{d}y^2 + \mathrm{d}z^2 \tag{1.36}$$

Recall that $w = -ict$. To bring us back to real time, we thus have ①

$$\mathrm{d}s^2 = -\mathrm{d}(ct)^2 + \mathrm{d}x^2 + \mathrm{d}y^2 + \mathrm{d}z^2 \tag{1.37}$$

The importance of $\mathrm{d}s^2$ is that, it is an invariant quantity under Lorentz transformation. To see that, one may first use $\{w, x, y, z\}$ and note that rotation keeps radius invariant. Or one can use the Lorentz transformation matrix to verify it explicitly. Just as (ω, x, y, z) forms a vector in 4d Euclidean space, $x^\mu = (ct, x, y, z)$ ($\mu = 0, 1, 2, 3$) also form a 4 dimensional vector (4-vector for short), which lives in the so called *Minkowski space* — which is just the mathematical name of our spacetime.②

In fact, "relativity" is not a good name to the theory. By its name, it appears to emphasize that things are relative — no longer as invariant as we thought.

But in fact, the essence of relativity is the other way round: Despite of being viewed from different perspectives (wrt different observers in different inertial frames), the events and their intrinsic relations are invariant.③

---
① In some books, the interval is defined as $\mathrm{d}s^2 = \mathrm{d}(ct)^2 - \mathrm{d}x^2 - \mathrm{d}y^2 - \mathrm{d}z^2$. This is just a different convention without physical difference (corresponding to replacing all $\mathrm{d}s^2$ here into $-\mathrm{d}s^2$).

② In fact, our spacetime is more complicated than the Minkowski space — in general relativity, you will see that our spacetime is curved instead of totally flat. This is considered to be more general spacetime but still with some Minkowski signature.

③ Recall the principles (ℝ), (ℂ) and (𝔼). Every each principle tell you a piece of invariance, instead of relativity.

Relativity is just the technology of study. But the spirit of the theory is the underlying invariance.

This is just like how you prove a theorem in Euclidean geometry — it is not important if the figure is placed vertically or horizontally (i.e. wrt observers with different viewing angles). What's important is the intrinsic relation between the geometrical objects.

Another analogue is the story of blind men and an elephant (Fig. 1.40). The goal of physics is to understand the elephant, instead of how different men feel differently by not being able to feel the whole thing.

Fig. 1.40

Time-like, space-like and null from the interval

The sign of ds^2 corresponds to different types of separations of events:

$$\begin{aligned} ds^2 < 0 &\Leftrightarrow \text{time-like} \\ ds^2 > 0 &\Leftrightarrow \text{space-like} \\ ds^2 = 0 &\Leftrightarrow \text{light-like, null} \end{aligned} \quad (1.38)$$

And now you know why null is called null — it is indeed null.

(Optional) A metric of spacetime

A metric is a "standard" to measure coordinate distance. For our spacetime in special relativity, the metric $g_{\mu\nu}$ (as a symmetric 4×4 matrix), infinitesimal coordinate distances and the invariant interval can be related as

$$ds^2 = \sum_{\mu,\nu=0}^{3} g_{\mu\nu} dx^\mu dx^\nu \quad (1.39)$$

where

$$x^\mu = \{ct, x, y, z\}, \qquad g_{\mu\nu} = \begin{pmatrix} -1 & 0 & 0 & 0 \\ 0 & 1 & 0 & 0 \\ 0 & 0 & 1 & 0 \\ 0 & 0 & 0 & 1 \end{pmatrix} \qquad (1.40)$$

This seems to be a trivial rewriting of what we have obtained. However, here we have separated two different things:

- Coordinates x^μ: What are the directions of spacetime?
- Geometry: What kind of geometry does spacetime satisfy?

Even if we use a different coordinate system (Fig. 1.41), for example, spherical coordinate for the spatial part, we can still choose a metric to keep ds^2 invariant:

$$x^\mu = \{ct, r, \theta, \phi\}, \qquad g_{\mu\nu} = \begin{pmatrix} -1 & 0 & 0 & 0 \\ 0 & 1 & 0 & 0 \\ 0 & 0 & r^2 & 0 \\ 0 & 0 & 0 & r^2 \sin^2\theta \end{pmatrix} \qquad (1.41)$$

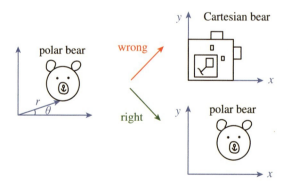

Fig. 1.41

In fact, the advantage brought about by a metric is not only the freedom to choose coordinates. Rather, it can describe the intrinsic curvature of spacetime which is not characterized by a different coordinate system. But this is beyond the scope of special relativity. You will see later that in general relativity, the metric $g_{\mu\nu}$ is the most important object to study. It can get curved by matter, and can guide matter how to move.

(Optional) Inner product of 4-vectors

The invariant internal ds^2 can be considered as the inner product of 4-vector dx^μ with itself. In general, we can use the metric to measure the inner product of any 4-vectors. For two 4-vectors p^μ and k^ν, their inner product $\sum_{\mu\nu} g_{\mu\nu} p^\mu k^\nu$ is Lorentz invariant (i.e. invariant under Lorentz transformation).

1.9 Relativistic Momentum and Energy

Before discussing energy and momentum in relativity, let us first recall why we need them at all. In Newtonian mechanics, it is nice to have energy and momentum because they are conserved. (See below for why conservation laws are good after all.) Thus in relativity, we should check if the Newtonian definition of energy and momentum are still conserved; and if not, how to generalize the definition of energy and momentum, to make them still conserved.

Transforming Slimes

Consider the following quiz:

Suppose there are slimes of 3 shapes: $\triangle$, $\square$ and $\bigcirc$. When slimes of differt shapes meet, they transform into the third shape. For example:

$$\triangle + \square \to \bigcirc \times 2$$

Now that we have $\triangle \times 5$, $\square \times 7$ and $\bigcirc \times 9$. Can you let them meet and transform them into $\triangle \times 6$, $\square \times 7$ and $\bigcirc \times 9$?

Obviously not. Since the total number of slimes is conserved when they meet.

Hope you get some feeling here that conserved quantities are helpful.

What's good about conservation laws?

(1) Physically, conservation laws are the magic that we do not need to care what's happening in the middle. Given the initial state, we can immediately make a number of predictions (no greater than the number of conservation laws in the system).

(2) Mathematically, the equations of motion (like a set of $m\ddot{x}+V'=0$) are second order ordinary differential equations (ODEs) in x which could be difficult. The conservation laws turns some of these into 1 or 0 order ODEs in x which are much simpler (e.g. $\frac{1}{2}m\dot{x}^2+V=E$, or charge conservation).

(3) Practically, in modern physics, the objects being studied are often extremely fast ($v \sim c$), small ($\Delta x \times \Delta p \sim \hbar$), heavy ($GM/(rc^2) \sim 1$), early (less than a second after the big bang), etc. It is often hard to directly build experiments to monitor exactly what is happening. Instead, one often needs conserved quantities to make useful observations. For examples, on modern colliders such as the Large Hadron Collider, the trajectories of particles (they are both small and fast in the above sense) are studied at a much later time compared to the time where interaction happened. If charge, energy or momentum were not conserved, it becomes much harder to study such particle physics processes (in fact, it will become not clear how to define observables even in principle without the help of energy and momentum conservation).

That's cool. But why are we still lucky to have energy and momentum conservation in special relativity? In a later chapter "From the Action to the Laws of Nature", we will show you a positive answer — the existence of these conservation laws are not a result of luck, but a result of the fact that no time moment or spatial location is special. This applies in special relativity as well as the Newtonian mechanics. Wait for that part if you are curious about the origin of these conservation laws.

1.9.1 Relativistic Momentum

Can we still use Newtonian momentum conservation?

First, let us recall how the Newtonian momentum conservation are consistent with (R) within Newtonian mechanics — forget relativity for a moment here and assume $v \ll c$.

For example, for Alice's momentum conservation for two particles:

$$m_1 \boldsymbol{v}_1^A + m_2 \boldsymbol{v}_2^A = m_1 \boldsymbol{v}_1'^A + m_2 \boldsymbol{v}_2'^A \tag{1.42}$$

What will Bob find? In Newtonian mechanics, mass of the particles do not change. And velocity simply adds. Thus for Bob with a relative velocity $\boldsymbol{v}$, $\boldsymbol{v}_1^B = \boldsymbol{v}_1^A + \boldsymbol{v}$, etc. Thus, equation (1.42) can be consistently written into the same form in Bob's frame:

$$m_1 \boldsymbol{v}_1^B + m_2 \boldsymbol{v}_2^B = m_1 \boldsymbol{v}_1'^B + m_2 \boldsymbol{v}_2'^B \qquad (1.43)$$

Happily, all the terms with $\boldsymbol{v}$ cancel in the above equation. So it takes the same form as Eq. (1.42). Thus, (ℝ) is indeed satisfied.

Now, let us take relativity into account. For simplicity, let us assert that the mass of the particles still does not change with velocity. *Assuming* that Alice still has the momentum conservation equation Eq. (1.42), then what will Bob find?

What if mass depends on velocity

Indeed, we can assume that mass depends on velocity. We will get essentially the same relativistic momentum, just with a different definition of mass, the so called "relativistic mass": $m_{\rm rel} = \gamma m$. We will not use this notation.

Instead, when we refer to mass, we mean the constant rest mass. Let the relative velocity between Alice and Bob be along the x direction as usual. Then in the x direction, Bob has:

$$m_1 \frac{(v_{1Ax} + v)}{\left(1 + \frac{v_{1Ax} v}{c^2}\right)} + \cdots = m_1 \frac{(v'_{1Ax} + v)}{\left(1 + \frac{v'_{1Ax} v}{c^2}\right)} + \cdots \qquad (1.44)$$

In the y direction (and z direction has a similar equation), Bob has:

$$m_1 \frac{v_{1Ay}}{\gamma \left(1 + \frac{v_{1Ax} v}{c^2}\right)} + \cdots = m_1 \frac{v'_{1Ay}}{\gamma \left(1 + \frac{v'_{1Ax} v}{c^2}\right)} + \cdots \qquad (1.45)$$

Unhappily, these equations *do not* take the same form as Eq. (1.42),[①] because of the red-colored factors. Thus, Eq. (1.42) takes different form wrt Alice and Bob. It cannot be a piece of law of nature in special relativity.

Now that we have seen the Newtonian energy and momentum does not lead to conservation laws in relativity (and the unexplained belief that there are still conserved energy and momentum after all),[②] the best we can do is to search for new quantities in special relativity, where in the $v \ll c$ limit the quantity returns to energy and momentum in the Newtonian mechanics.

---
　① The same argument applies to the Newtonian energy.
　② In fact, we only need the energy and momentum *differences* to return to Newtonian mechanics when $v \ll c$. You will find a surprise about energy in this respect soon.

Using a proper time to fix momentum conservation

How to fix the inconsistency of Eq. (1.44) and Eq. (1.45) with (ℝ)?

If we can remove the highlighted factor, we should be able to get a equation satisfying (ℝ). How? We recall that it arises because

$$dt_B = d\left[\gamma\left(t_A + \frac{x_A v}{c^2}\right)\right] = \gamma(1 + \frac{v_{1Ax}v}{c^2})dt_A \tag{1.46}$$

In other words, every observer has their own time. Measuring motion with individual observer's own time ($d(\cdots)/dt_A$ and $d(\cdots)/dt_B$) is the root of evil.

How can we get rid of the observer's own time in defining momentum? Or, is there an observer-independent way to measure a time-like interval?

As I am trying to rephrase the question to approach the answer, now you should be able to figure it out: we have learned an observer-independent way to measure intervals using ds^2. To make a real number with correct time dimension, we define the *proper time*:

$$d\tau \equiv \sqrt{-ds^2/c^2} = \sqrt{dt^2 - \frac{dx^2}{c^2}} = \frac{dt}{\gamma} \tag{1.47}$$

This is the time variable that we need.

Physical meaning of proper time

Why a time variable can be defined *independently of* an observer? Haven't we said that there is no absolute time?

Because there exists a moving object (it moves, therefore it is). The moving object defines its own frame and thus defines its own time. In the own frame of the moving object, $d\boldsymbol{x} = 0$. Thus indeed $d\tau = dt$. This verifies that the proper time is the time measured by the moving object itself. Imagine a clock moving together with the object. The proper time is the interval read from this clock.

Thus, the momentum which gives a relativistic generalization of momentum conservation is

$$\boldsymbol{p} = m\frac{d\boldsymbol{x}}{d\tau} = \gamma m \boldsymbol{v} \tag{1.48}$$

From the first equal sign of Eq. (1.48), one observes that $\boldsymbol{p}$ transforms in the same way as $\boldsymbol{x}$ under Lorentz transformation. This is as promised that momentum conservation takes the same form in different frames.

However, this is confusing from another aspect: we know that space ($\boldsymbol{x}$) and time (ct) has combined into spacetime ($ct, \boldsymbol{x}$) in relativity as one entity. Now that momentum transform the same as space, is there a counterpart comes together which transform the same as time (ct)? What do we get if we naively replace $\boldsymbol{x}$ with ct in Eq. (1.48)? We get

$$\text{(the naive time-like counterpart of } \boldsymbol{p}) = m\frac{\mathrm{d}(ct)}{\mathrm{d}\tau} = \gamma mc \tag{1.49}$$

What's this? The nature should not leave it unexplained! We will return to this question in the next subsection.

Caution: Here we have only showed that the definition (1.48) is consistent with (ℝ). We have not proved for you the actual conservation. In Newtonian mechanics, momentum conservation is derived from the 3rd law (which is an independent assumption from the Newtonian 1st and 2nd laws). Here we did not impose a 3rd law (although we can do so) and thus we do not have the right tool to derive momentum conservation. In fact, given an action, the relativistic momentum conservation can be derived from an action principle. We will not expand this point here.

The relativistic force

Now that there is momentum, it is straightforward to define force:[①]

$$\boldsymbol{F} \equiv \frac{\mathrm{d}\boldsymbol{p}}{\mathrm{d}t} = \frac{\mathrm{d}(\gamma m \boldsymbol{v})}{\mathrm{d}t} = \gamma m \dot{\boldsymbol{v}} + \gamma^3 m \boldsymbol{v}(\boldsymbol{v} \cdot \dot{\boldsymbol{v}})/c^2 \tag{1.50}$$

Note that when $v \to c$, it takes $F \to \infty$ to change v. As a result, one can never accelerate a subluminal object to speed of light (or beyond). Thus c is indeed the speed limit.

> **Chasing the light?**
>
> When Einstein was 16 years old, he started to dream what would happen if he can run as fast as light. Would light stop oscillating and would that contradict Maxwell's theory of E&M?
>
> Now his dream has came to an end by his own efforts. No one can be accelerated to the speed of light and thus this seemingly contradictive thought experiment would never happen.

[①] Here we use dt instead of dτ. This definition is convenient, because it is intuitive to think of force as one static observer pushes an object and see its acceleration — how much additional efforts the static observer has to consume to push the rocket further.

1.9.2 Relativistic Energy

The relativistic kinetic energy

The relativistic force implies work to accelerate objects, and thus the kinetic energy. Consider an object initially at rest at location $x = 0$, and is accelerated by a force in the x-direction, into a state with position x and speed v. The work that the force has done turns into the kinetic energy of the object:

$$K = \int_0^x F \, \mathrm{d}x = \int_0^x \frac{\mathrm{d}(\gamma m v)}{\mathrm{d}t} \, \mathrm{d}x = \int_0^v v \, \mathrm{d}(\gamma m v) = (\gamma - 1) m c^2 \qquad (1.51)$$

In the second-to-last step we have converted the integration variable to v, and at the final step, we have performed integration by parts. The detail is left as an exercise.

A few comments in order before to proceed:

(1) When $v \ll c$ (so that we can do a Taylor expansion around $v \to 0$), the relativistic kinetic energy returns to our familiar Newtonian form as expected: $K \to \frac{1}{2} m v^2$.

(2) We have discussed kinetic energy following an acceleration process in the x-direction from rest. Is it general enough? Since kinetic energy is a label of a state, it should not care how it is obtained in the history of the object. Thus the kinetic energy is general enough no matter the force is not in the x-direction, or the object has an initial velocity.

Now that we have the kinetic energy, what's the total energy? Before that, let us first study mass. In Newtonian mechanics (and chemistry), mass is conserved — the total mass before and after a reaction is the same. What about in relativity?

Mass is no longer conserved

We study a system split into two objects. Consider two objects connected by a spring (Fig. 1.42). The spring is compressed and stores some potential energy. Initially the objects are at rest. The initial mass of the whole system is M. After releasing the spring, the two objects move apart, each has mass m and moving apart with speed v.

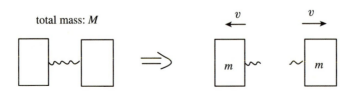

Fig. 1.42

In Newtonian mechanics, there is no doubt that $M = 2m$. But is it true when $v \sim c$?

To study the property of the system, we add a small probe velocity in the y direction $v_y \ll v$ (One can comfortably take v_y to be infinitesimal). Then interestingly, momentum conservation in the y direction can calculate M for us.

(1) Before the split: $p_y = Mv_y$.

(2) After the split: $p_y = 2\gamma m v_y$ for the same p_y (momentum conservation), where $\gamma = 1/\sqrt{1 - v^2/c^2}$ accounts not only velocity in the y direction, but the the total velocity.

Thus, $M = 2\gamma m > 2m$! The good old mass conservation breaks down.

> **Does v_y change?**
>
> No. Because we can view the event in another frame with $v_y = 0$ at the beginning. Then in this frame $v_y = 0$ at the end. Now switch to a frame with a small $v_y \ll c$. Using the velocity addition rule, v_y does not change. (And in the x direction v_x only get a small correction of order $1/\sqrt{1 - v_y^2/c^2} - 1 = \mathcal{O}(v_y^2/c^2)$.)

The relativistic rest energy

Consider the above $M \to 2m$ split process. Let's consider the limit $\gamma \gg 1$, i.e. the final objects are flying very close to the speed of light. In this limit, the kinetic energy of the final objects are $K \simeq 2\gamma m c^2$.

From energy conservation, the kinetic energy of the final objects is contained in the initial object. Thus, the energy of the initial object is

$$E = 2\gamma mc^2 = Mc^2 \qquad (1.52)$$

Recall that the initial object is at rest. Thus, the energy here is the rest energy of an object.

---

① The two final objects also has their rest energies, mc^2 each. In our $\gamma \gg 1$ limit, these rest energies can be neglected. In fact, initially Einstein used light to derive the relativistic energy, corresponding to no rest mass. But the derivation needs more understanding of E&M than assumed in this course. We thus follow another route here.

Daily energy usage of Hong Kong

To see the huge rest energy of matter, for example, the daily energy consumption of Hong Kong is of order 10^{15} J. That corresponds to about 10 g of matter — about the weight of one AAA battery. To compare, the chemical energy that an AAA battery can provide is a few thousand joules.

Why do stars shine?

Before relativity was understood, there was a big mystery that how stars can shine for longer than the human history, if the stars are powered by chemical energy. Instead, the nuclear fusion in the star can power the stars for billions of years.

The rest energy

$$E_{\rm rest} = mc^2 \qquad (1.53)$$

is the part of energy that an object has, even if it is not moving at all. This energy is huge in our daily standard since c is a huge number in our daily life (and we have $c \times c$).

In Newtonian mechanics, due to mass conservation, this energy is not noticed in energy conservation. However, in relativistic situations this energy can be released. For example,

(1) We have already encountered a non-trivial case of rest energy above — the two final objects after the split having $E = Mc^2 > 2mc^2$.

(2) In more realistic situations, the usage of nuclear energy makes use of the rest energy of matter. In nuclear fissions and fusions, a much greater amount of energy can be released compared to usual chemical or mechanical reactions.

(3) In labs, one can collide particles and anti-particles. The particles disappear after they meet and the rest energy can be completely released in the form of light.

The zero-point of rest energy

The particle-anti-particle annihilation phenomenon is a very important check about the concept of rest energy. Because the zero point of energy makes no physical sense unless all the energy can be released.

The relativistic energy

In general, for an object with mass m and speed v, combining the kinetic energy (1.51) and the rest energy (1.53), the total energy of the object is

$$E = \gamma m c^2 \tag{1.54}$$

This sounds a bit different from what you heard: $E = mc^2$. This more famous formula (famous because without having to explain γ in popular science) may stand for one of the two meanings:

(1) The rest energy of an object.

(2) The total energy of an object, with the mass defined as the velocity-dependent "relativistic mass" $m_{\text{rel}} \equiv \gamma m$ and thus $E = m_{\text{rel}} c^2$. Nowadays we understand mass as a Lorentz invariant label of a particle; the convention of m_{rel} is seldomly used and thus we will not use m_{rel} further here.

The 4-dimensional momentum vector

Remember the question we asked at Eq. (1.49)? Happily, up to a factor of c (which is nothing more than a conversion of unit as we have explained natural unit), the energy we have got in Eq. (1.54) is exactly the missing "naive time-like counterpart of momentum" in Eq. (1.49). The 4 dimensional momentum is then $p^\mu = (E/c, \boldsymbol{p}) = (\gamma m c, \gamma m \boldsymbol{v}) = \gamma m \dot{x}^\mu$.

Recall that for the spacetime coordinate 4-vector, we have an invariant quantity $c^2 \Delta t^2 - \Delta x^2$. Is there a counterpart for the momentum 4-vector? Yes. Squaring Eq. (1.54), we get

$$m^2 c^4 = E^2 - \frac{v^2}{c^2} E^2 \xrightarrow{\text{using ratio of Eq. (1.54) and Eq. (1.48)}} E^2 - p^2 c^2 \tag{1.55}$$

Thus, though E and $\boldsymbol{p}$ are dependent on observers, the combination on the RHS is independent of observers, but is just the invariant mass squared of the particle. This is not surprising. Because the momentum and coordinate 4-vectors lives in the same space and is measured by the same metric. (Just as it is not surprising that both the magnitude of 3d coordinate vector and 3d momentum are rotational invariant.)

> **Light is light**
>
> Applying Eq. (1.55) to light, we have $v = c$. Thus, $m = 0$. In other words, if we view light as particles (see the quantum mechanics part for more information), their mass has to vanish.

The 4-vector of the momentum of light has the form $p^\mu = (E/c, \boldsymbol{p})$, where the energy of light $E = c|\boldsymbol{p}|$. This 4-vector has vanishing Lorentz norm. In the notation of the metric (optional material), we have $\sum_{\mu,\nu} g_{\mu\nu} p^\mu p^\nu = 0$.

(Optional) The relativistic Doppler effect of light

Alice is moving with 3-velocity $\boldsymbol{v}$ wrt Bob. If Bob observes that the energy of light is E_B, what's the corresponding energy E_A of the same beam of light in Alice's frame?

We can solve this problem by the Lorentz invariance of inner products of 4-vectors[1]. The inner product $\sum_{\mu\nu} g_{\mu\nu} k^\mu p^\nu$ is invariant under Lorentz transformations, where k^μ and p^ν are the 4-vector of Alice's momentum and the momentum of the beam of light.

In Bob's frame: $k^\mu = (\gamma mc, \gamma m\boldsymbol{v})$, $p^\nu = (E_B/c, \boldsymbol{p})$. The Lorentz invariant inner product is $\sum_{\mu\nu} g_{\mu\nu} k^\mu p^\nu = -\gamma m(E_B - \boldsymbol{v}\cdot\boldsymbol{p}) = -\gamma m E_B(1 - (v/c)\cos\theta)$, where θ is the angle between $\boldsymbol{v}$ and $\boldsymbol{p}$ in Bob's frame.

In Alice's frame: $k^\mu = (mc, \boldsymbol{0})$, $p^\nu = (E_A/c, \boldsymbol{p}')$. The Lorentz invariant inner product is $\sum_{\mu\nu} g_{\mu\nu} k^\mu p^\nu = -m E_A$. It must be equal to the Lorentz invariant in Bob's frame. Thus, $E_A = \gamma E_B(1 - (v/c)\cos\theta)$.

From either classical E&M or quantum mechanics, we will find that the frequency of light is $\omega \propto E$. Thus the frequency of light wrt Alice and Bob has a similar relation $\omega_A = \gamma \omega_B(1 - (u/c)\cos\theta)$.

1.10 Epilogue: Summary and What's Next

The below diagram is a recap of what we have learned (Fig. 1.43).

---

[1] Here the purpose of this optional box is not only to tell you the formula of the relativistic Doppler effect, but also to show the power of Lorentz invariance for the inner product. The Doppler effect can also be derived just by considering the effect of time dilation and space contraction. But the calculation is much more complicated for a general direction θ.

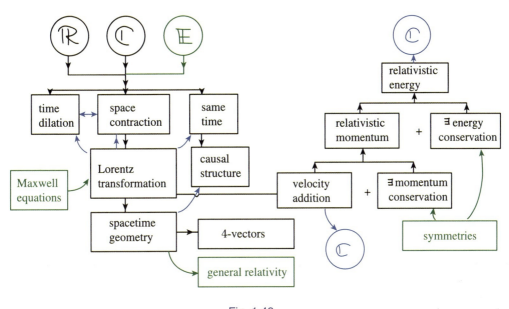

Fig. 1.43

Further reading

(1) If you are interested to explore more about the geometry of spacetime and formulate special relativity with geometrical emphasize, read "Spacetime Physics" by Taylor and Wheeler (and a few general relativity books below).

(2) If you are not happy about our guess of relativistic energy and momentum, but rather want to derive them, wait and we will do it in the part "From Action to Laws of Nature". See also "Theoretical Minimum" (Video Lectures) by Susskind.

(3) If you consider the math used this text too sloppy, read "Relativity: Special, General, and Cosmological" by Rindler.

(4) If you like to learn more about the relation between electrodynamics and relativity before getting to the full details of electrodynamics, read "Lecture Notes on Modern Physics" by Baumann.

(5) Many good books on general relativity also starts with a dense and high-level introduction of special relativity, for example "Gravity: An Introduction to Einstein's General Relativity" by Hartle, "Gravitation" by Misner, Thorne and Wheeler and "Gravitation And Cosmology" by Weinberg.

What happens next in a university physics program?

(1) Electrodynamics as a deeper study of E&M.

When Einstein was young, he was deeply puzzled by two observations. Both has roots in E&M. One is light cheasing, which you have understood by now. The other was the magnet-conductor paradox, as he emphasized in the first paragraph of his 1905 paper. The paradox is shown in Fig. 1.44.

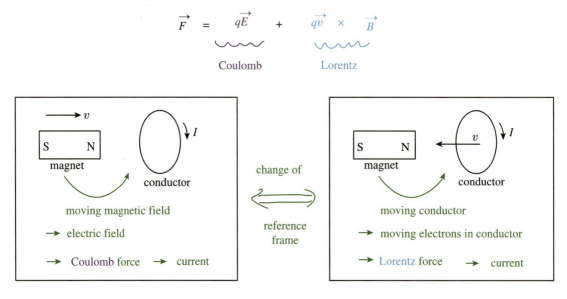

Fig. 1.44

Before special relativity, the Coulomb law and the Lorentz law are considered as two independent fundamental laws of nature. However, by a change of frame, in the moving-magnet frame, the induct current is explained using the Coulomb law and in the moving-conductor frame the induced current is explained using the Lorentz law. This indicates that one should not consider them both fundamental — one same thing shouldn't be explained by two fundamental principles in physics! If the Coulomb law is more fundamental (at least it is more familiar), we should be able to derive the Lorentz law from it.

How relativity helps? Roughly speaking, 3-dimensional vectors such as E and B should be extended into 4-dimension vectors. Thus, the Coulomb and Lorentz laws, depending only on E and B, respectively, should be combined into one law using the form of 4-velocity and 4-electric-magnetic vector.

Without getting into the math, let us use a thought experiment to intuitively understand how to "derive" Lorentz force from Coulomb force by a change of frame.

Consider a wire conducting electric current, and a charge. The wire and the charge has relative motion between each other (Fig. 1.45).

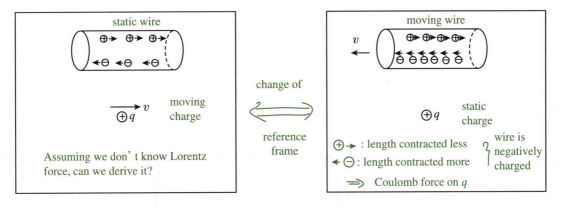

Fig. 1.45

Here the Lorentz force is derived by the different amount of length contraction effect of moving charges.

In electrodynamics, you will explore the full connection between E&M and relativity.

(2) You may learn how special relativity works with gravity in "general relativity". We will also have a part to mention it briefly.

(3) Symmetry, transformation and group. Einstein's postulates (to be more precise, the Lorentz transformation) has the mathematical structures of a group (Lorentz group). "Group theory" is widely used in physics, including relativity, particle physics and solid state physics. And they are of their own importance in math as well.

(4) You may learn how special relativity works with quantum mechanics in "quantum field theory".

Exercises

E1.1 Plain waves

Consider "plain wave" $e^{ik(x-ct)}$. Interpret the real part of $e^{ik(x-ct)}$ as the amplitude of the wave.

(1) For fixed t, show the plain wave indeed looks like a wave.

(2) Figure out the moving direction and speed of the wave.

E1.2 Light travel and time dilation

Consider Alice and Bob have relative motion against each other, with velocity v (Fig. 1.46). Alice carries a candle, which emits light (speed of light is c) perpendicular to the motion direction (wrt Alice).

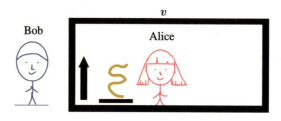

Fig. 1.46

(1) What's the speed of this light ray wrt Bob?
(2) What's the travel direction of this light ray wrt Bob?
(3) Wrt Bob, the time of Alice slows down. Why doesn't the speed of this light ray slow down? Explain the relation between slower time and speed of light.

E1.3 Spacetime diagram of the sun-earth system

Use the frame that the sun is static, draw a spacetime diagram of the sun's and the earth's motion in x-direction (a direction in the earth's orbit plane, Fig. 1.47). Show on this diagram how the sunlight (emitted in the x-direction) reaches the earth. The vertical axis is ct.

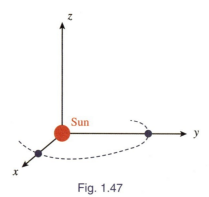

Fig. 1.47

E1.4 Spacetime diagram with constant proper acceleration

Alice is moving with constant "proper acceleration" in the x-direction: $x^2 - t^2 =$ constant (to simplify the discussion, let's work in one spatial dimension only). Not all light towards the observer can reach Alice. The region where light cannot reach Alice is called the "Rindler horizon". Draw a spacetime diagram of Alice's motion and find where the Rindler horizon is.

E1.5 What the twin actually sees

Draw the spacetime diagram of the twin paradox and show what the static twin actually sees (i.e. in the order that light reaches his eyes) about the aging of the moving twin (when she was moving outwards and when she was moving back).

E1.6 Distance between spaceships

An observer A is considered at rest in the whole setup of this question. Two spaceships B and C are equidistant to A and are initially also at rest, and the distance between them is L. Now A sends a light signal. After receiving the light signal, B and C immediately start to move at a velocity v in the same direction (neglect the period of acceleration). What is the distance between B and C after they are moving? Give your answer wrt A and wrt B, respectively. Hint: draw a spacetime diagram to find out what happened.

E1.7 Use Lorentz transformation in calculations

Use Lorentz transformation to calculate:
(1) Time dilation.
(2) Ruler contraction.
(3) For two events which happened at the same time wrt Alice, calculate the time difference wrt Bob, given Bob's speed v, and the distance between the two events being L wrt Alice.

E1.8 Examples of velocity addition

Alice is moving away from Bob with velocity $\boldsymbol{v} = (v,0,0)$, and sending out light rays.

(1) If the light ray is along x_A direction (x direction wrt Alice), calculate the velocity of the light ray $\boldsymbol{v}_B = (v_{Bx}, v_{By}, v_{Bz})$ wrt Bob.

(2) If the light ray is along y_A direction (y direction wrt Alice), calculate the velocity of the light ray $\boldsymbol{v}_B = (v_{Bx}, v_{By}, v_{Bz})$ wrt Bob.

E1.9 Speed of light in the media

Consider the speed of light in static air (refractive index $n = 1.0003$). How does the speed of light in the air change wrt moving observers moving with speed v? How does the speed of light in the air change when there is a wind with speed v?

E1.10 The spacetime interval is Lorentz invariant

Show that under Lorentz transformation, the spacetime interval $\mathrm{d}s^2$ is unchanged (although t and x change). For simplicity, work in two dimensions t and x only (i.e. no motion or rotation in y and z directions).

E1.11 Our motion in spacetime

Show that using proper time, everyone is moving in spacetime (not space) with the same 4-speed (i.e. the size of the 4-velocity $\mathrm{d}x^\mu/\mathrm{d}\tau$).

E1.12 Integrate momentum to get energy

Let us use the relativistic momentum $p = \gamma m v$ to derive the expression of the kinetic energy, in a different way from what we did in class.

Consider a ball at rest at $x = 0$ with mass m, and act a constant force F on this ball toward the x direction. The ball then accelerates because of the force.

Note that $F = \mathrm{d}p/\mathrm{d}t$, and the kinetic energy can be calculated from the work done by the force:

$$K = \int_0^{x_1} F \, \mathrm{d}x = \int_0^{x_1} \left(\frac{\mathrm{d}p}{\mathrm{d}t}\right) \mathrm{d}x = \int_0^{p_1} v \, \mathrm{d}p = \int_0^{v_1} v \, \mathrm{d}(\gamma m v) \tag{1.56}$$

where x_1, p_1 and v_1 are the distance, momentum and velocity at a later time t_1.

Continue the calculation and derive the kinetic energy of the ball at time t_1.

E1.13 Read Einstein's original papers on relativity

Nowadays Einstein's original papers can be easily found online. For example, his first paper on special relativity. You will find most parts of the paper accessible except that in electrodynamics he used different notations from modern convention.

Chapter 2
General Relativity

The earth has gravity. As a result ——

① Your upstairs neighbor gets older faster than you.
② A person standing in front of you is shorter than he looks.
③ Consider a triangle, whose edges are shortest distance lines. The sum of its inner angles is greater than 180° if you hold it horizontally; and smaller than 180° if you hold it vertically.

These effects are too small to notice, since the earth gravity is weak. If the earth was extremely massive, not only the above effects are more dramatic, but also

④ The earth turns black. Also, the earth can become much brighter than now.
⑤ If you get closer, the center of the earth becomes no longer anywhere in space. It becomes your future.

All of these are due to gravity. But let me tell you a secret—there is in fact no gravity.

2.1 The Equivalence Principle

Dropping a feather and a stone, by you and by Newton

If you drop a feather and a stone in the vacuum, they fall at the same acceleration. Are you surprised by this fact?

Perhaps you were surprised when you were a kid, but no more after you have learned Newtonian mechanics—it's trivial:

$$\begin{aligned}\text{Newtonian 2nd law: } F &= ma. \\ \text{Newtonian gravity: } F &= mg.\end{aligned} \quad \Rightarrow \quad a = g \text{ for all materials.}$$

But Newton himself did not find it trivial. In Principia, he wrote:

> It (mass defined by F/a) can also be known from a body's weight, for—by making very accurate experiments with pendulums[①]—I have found it to be proportional to the weight... I have tested this with gold, silver, lead, glass, sand, common salt, wood, water, and wheat.

Why Newton was so careful here?

Why two quantities are one?

To see why Newton is so careful, let us examine the meaning of mass more carefully. We can define two quantities:

Inertial mass: $m_\text{I} = F/a$ is a measure of inertia—laziness in changing its velocity.

Gravitational mass: $m_\text{G} = F/g$ for gravity—how strongly gravity attracts the matter.

Thus, m_I and m_G are from different origin. However, Newton asserts that $m_I = m_\text{G}$. And state-of-art experiments tell that $|(m_\text{I}/m_\text{G}) - 1| < 10^{-16}$. Instead of accepting $m_\text{I} = m_\text{G}$ as an experimental rule (actually, infinite number of experimental rules—for gold, silver, lead, glass, ..., respectively), the curious would like to ask:

Is there a fundamental explanation for the equivalence $m_\text{I} = m_\text{G}$?

Why gravity is so special, while other forces have no such equivalence?

Gravity is special

Gravity is special here. Force is not always proportional to mass. Electric force $F = qE$ is proportional to the electric charge of matter, instead of mass. So for electric force, the strength of force is determined by charge and inertia is determined by mass. Their ratio differs for different kinds of matter. But for gravity, both strength of force and inertia are proportional to one quantity—mass.

---

① Pendulum experiment is a smarter way to test how things fall, since it's slower and periodic.

These questions are answered by Einstein. Guess how did he answer them? In 1907, the "happiest thought" in Einstein's life arrives—"If a person falls freely, he will not feel his own weight." Let's explore the power of this thought.

Einstein's equivalence principle

In 1907, Einstein asserts that, for uniform gravitational field:

"We... assume the complete physical equivalence of a (uniform) gravitational field and a corresponding acceleration of the reference frame. "

For example, if Alice is in a lift and does not look outside (Fig. 2.1). Then she finds no difference between the following situations:

Fig. 2.1

(1) The lift with constant acceleration a in an environment without gravity.

(2) The lift is not moving, but placed in a gravitational field $g = -a$.

We added "uniform" to Einstein's words to make it more precise. If a gravitational field is not uniform, we can always study a small enough element of space, where the gravitational field is uniform. (How to "patch" these small elements together is non-trivial and will result in curved spacetime.)

Why this assumption

Everybody can make assumptions. Why is this one special? The assumption is two-fold:

Special relativity restricts our attention to inertial frames. The accelerated frame broadens the scope of relativity.

Gravity gets downgraded from a "regular" force to something fundamentally equivalent to a fictitious inertial force.

This assumption indeed answered the questions: $m_I = m_G$ because inertial force (acceleration) equals gravity. And gravity is special because this is the force in this fundamental assumption.

Here we name a few simple consequences of the assumptions, and leave more surprising ones to later sections as they need more explanations.

Uniform gravity can be cancelled by constant acceleration

This is because uniform gravity g is equivalent to constant acceleration $a = -g$, and constant acceleration $-g$ can be cancelled by constant acceleration g.

For example, one feels no gravity in free falling lifts, or satellites orbiting around the earth. These observers can in fact consider themselves as inertial frames. For these observers, the uniform gravity field can be considered non-existent (as it cannot be probed by any experiment).

Thus uniform gravity is indeed special (compared to all other forces such as E&M)—it is as fictitious as inertial force. Its existence (or not) is observer dependent.

Light bends in uniform gravity

How light move in a uniform gravitational field? This problem may be considered as not well-defined in Newtonian mechanics as the mass of light is zero. But with equivalence principle, we immediately know that light bends the same way as in an accelerated environment. This explains item ② at the beginning of this part.

> **Stars, light bending and lensing**
>
> For a similar reason, light also bends when travelling near an object, such as a star, or a galaxy. One cannot compute the bending angle only from the equivalence principle since the gravitational field from a star is not uniform. From a calculation in general relativity, the bending angle is twice of the angle naively calculated from the equivalence principle. This bending angle is the first verified prediction of general relativity (1915).
>
> To see the light bending effect in another way: objects focus light like what a lense can do. This effect is known as gravitational lensing. Gravitational lensing is well observed and has become a powerful tool in understanding the matter distribution of our universe.

2.2 Time with Uniform Gravity

Recall that in the twin paradox, acceleration makes the physical time difference between Alice and Bob (as Alice has to accelerate to return to meet Bob again). Now acceleration equals uniform gravity, we expect that gravity should cause some time differences as well between observers. We will see that it is indeed the case.

Higher is faster

Now Alice stands higher by a vertical distance h than Bob. The gravitational acceleration is g. Does Alice's time lapse differently from that of Bob?

To quantify "time lapse", consider that Alice sends two light pulses with interval Δt_A. What will be the time interval of the two pulses when Bob receives them?

For computational simplicity, let's assume $c\Delta t_A \ll h$ and $gh \ll c^2$. The situation is illustrated in the left panel of the below figure.

To solve this problem, we make use of the equivalence principle. The system is equivalent to the situation in the right panel of Fig. 2.2: There is no gravity. But rather Alice and Bob are accelerating upwards with acceleration $a = g$.

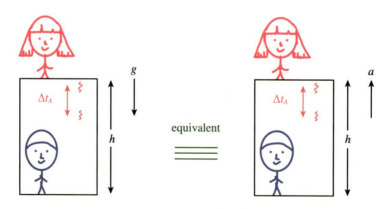

Fig. 2.2

We let $t = 0$ be the time that Alice sends the first signal, and also the time that Alice and Bob has velocity $v = 0$. Let the spatial separation of the light signals be s. Note that s should be a constant because there is no gravity or other forces affecting the propagation of light.

Wrt Alice, when she sends the signal, her velocity is $v = 0$ and thus $s = c\Delta t_A$.

Wrt an observer static at Bob's location, when Bob receives the signal, his velocity is $v = ah/c$. Note that the light signals and Bob are moving towards each other. Thus $s = (c+v)\Delta t_B$. Thus

$$\Delta t_A = \left(1 + \frac{ah}{c^2}\right)\Delta t_B \tag{2.1}$$

Thus, Bob finds Alice faster. The higher, the faster. If at $t=0$ Alice and Bob has the same age, later Alice will find Bob younger and Bob will find Alice older.

This explains item ① at the beginning of this part.

> **The twin paradox again**
>
> Now we are allowed to use the traveling twin's frame to interpret the twin paradox. The traveling twin needs to accelerate before returning. If we model the acceleration to be uniform acceleration, then that corresponds to a gravitational potential due to the equivalence principle. And the static twin suddenly becomes old in this gravitational potential.

Time dilation with a gravitational potential

More generally, for Alice and Bob in a gravitational potential ϕ, the time dilation can be expressed in terms of their gravitational potential difference as

$$\mathrm{d}t_A = \left(1 + \frac{\phi_A - \phi_B}{c^2}\right)\mathrm{d}t_B \tag{2.2}$$

The metric for a spherical star

Consider a spherical star with mass M. Let Bob be at $r \to \infty$, i.e. far away from the star. Then Bob does not experience gravity from the star, and has $\phi_B = 0$. We can thus regard Bob's time as standard time without the star gravity: $t_B = t$.

---

① We employed a 3rd observer C static at Bob's location to study Bob's motion wrt light. To talk about what is Bob's time interval, in fact there is time dilation effect between Bob and C. However the time dilation effect is of order v^2/c^2 and thus small in this analysis.

While Alice has gravitational potential $\phi_A = -GM/r$. Using Eq. (2.2), we get[1]

$$dt_A = \left(1 + \frac{\phi_A}{c^2}\right) dt \qquad (2.3)$$

For Alice static at a fixed position wrt the star, $d\boldsymbol{x} = 0$, and

$$ds^2 = -c^2(\text{proper time})^2 = -c^2 dt_A^2 = -\left(1 + \frac{2\phi_A}{c^2}\right) c^2 dt^2 \qquad (2.4)$$

In general, the coefficient in front of $d\boldsymbol{x}^2$ (i.e. the spatial part of the metric) is also modified. We will only give the result here:

$$ds^2 = -\left(1 - \frac{r_S}{r}\right) c^2 dt^2 + \left(1 - \frac{r_S}{r}\right)^{-1} dr^2 + r^2 \left(d\theta^2 + \sin^2\theta d\phi^2\right) \qquad (2.5)$$

This is known as the Schwarzschild metric, describing the spacetime geometry outside a spherical symmetric star. The radius

$$r_S \equiv \frac{2GM}{c^2} \qquad (2.6)$$

is called the Schwarzschild radius.

Example: The sun

To put in numbers for Eq. (2.6), if M is the solar mass, then $r_S = 3$ km. What does this 3 km distance mean to the sun?

It does not mean anything. Because the solar radius is $r_\odot = 7 \times 10^5$ km. The Schwarzschild metric (2.5) only applies outside the star, since inside the star, the gravitational potential takes a different form.

However, what if an object is so dense, that its radius is smaller than r_S? For example, what if an object has the same mass of the sun, but with a radius smaller than 3 km?

---

[1] Note that the gravitational field by the star is not uniform. But fortunately Eq. (2.2) can still be used here. And this even holds when gravity is very strong. The explanation (solving Einstein equations for the star) is beyond the scope of this short introduction.

2.3 Black Holes

Now imagine an object being so dense that its size is smaller than its Schwarzschild radius $r_S = 2GM/c^2$. What happens?

In fact, this is not a question living in imagination. Such dense objects—black holes—are real. They are detected from many observational methods, from the analysis of celestial motion to imaging its shape, to the observation of gravitational waves.

What happens near $r = r_S$?

When $r \approx r_S$, in Eq. (2.5) the time part vanishes and the spatial part blows up. What's happening there?

Imagine that Alice is staying at a distance r very slightly greater than r_S. Gravity pulls her to fall, but imagine that, she is in a powerful rocket to keep her staying at a fixed r. And Bob locates at $r \to \infty$. For any finite interval Δt_A according to Alice, wrt Bob:

$$\Delta t_B = \Delta t_A \times \left(1 - \frac{r_S}{r}\right)^{-\frac{1}{2}} \to \infty \tag{2.7}$$

> **Duality and holography**
>
> Similar to the twin paradox, now Alice and Bob has a disagreement. Alice sees herself falling across r_S but Bob sees Alice frozen at r_S. We will explain that they do not have a chance to meet (up to quantum subtleties which remain open questions in research) and thus nothing goes wrong.
>
> Bob sees and describes Alice on the 2-dimensional surface $r = r_S$, while Alice sees and describes herself in a 3-dimensional volume $r < r_S$. The two descriptions may be equivalent to each other. This conjecture is known as holography in theoretical physics as the dual descriptions are between a surface and a volume.

Thus, Bob finds Alice frozen on the $r = r_S$ surface. In general, at $r \to r_S$, the gravitational dilation effect is so significant that time looks frozen wrt an outside observer. In this respect, if you travel close to $r \to r_S$ and then back, it's a time machine that brings you to the future.

The above is what Bob finds. But wrt Alice, she does not find $r = r_S$ very special. As at any moment, assuming that she is much smaller than r_S, equivalence principle tells that her acceleration cancels her gravity and she just crosses $r = r_S$ from outside to inside with finite time.

An "event horizon" and a "black hole"

What will Bob see for the events with $r < r_S$? For example, after Alice crosses r_S to reach $r < r_S$, what can Bob see about Alice?

Nothing. Due to the extreme time dilation, the light emitted at r_S need infinite time to reach Bob. Bob does not have more than infinite time to wait and thus cannot see anything happening at $r < r_S$.

> **Are black holes really black?**
>
> No.
>
> Theoretically, Hawking radiations from quantum gravity can come out of black holes. They are usually too dim to see.
>
> Practically, some black holes are the brightest objects in our universe—seriously. Matter falling into the black hole (infinitely deep gravitational potential) interact with each other and emit bright radiations—up to 40% of matter energy can be converted to light. Due to the emission of infalling matter, images of black holes have been taken by arrays of radio telescopes in 2019.
>
>

As a result, $r = r_S$ is a surface to limit the events that Bob can see. Thus, $r = r_S$ is known as an event horizon. The dense object hides inside this event horizon and thus is invisible. No light from this object can reach to outside. Thus, the object is known as a black hole. This explains item ④ at the beginning of this part.

"Inside" the horizon: the future is doomed at a singularity

What happens after Alice falls inside a black hole (cross the $r = r_S$ horizon)?

Wrt Bob, he cannot expect to see Alice coming out again. Since when she crosses the horizon, it already corresponds to $t_B \to \infty$. Bob can see nothing happening later than infinity.

But wrt Alice herself, why cannot she decide to power her rocket to escape the horizon?

To answer this question, let us take a look at the Schwarzschild metric (2.5). At $r < r_S$, the metric can be rewritten as

$$ds^2 = -\left(\frac{r_S}{r} - 1\right)^{-1} dr^2 + \left(\frac{r_S}{r} - 1\right) c^2 dt^2 + r^2 \left(d\theta^2 + \sin^2\theta d\phi^2\right) \tag{2.8}$$

What happened? Now dr is time-like and dt is space-like! The time direction and one space direction have flipped their roles. For Alice, the $-r$ direction is now the time direction.[①]

The time direction is special since it is one-way. You can choose to move left or right in space, and have no choice but moving forward in time. Now Alice has no choice other than to move along her time direction (the direction to reduce r) towards $r = 0$, regardless of the spatial direction (now t, θ, ϕ) that Alice would like to travel.

Not only Alice, but everything inside the black hole, including the dense object itself, falls towards the future at $r = 0$. Their time ends here and thus it is known as a singularity. In fact, following calculations of general relativity, before reaching the singularity, every object is teared apart due to diverging tidal force near the singularity.

Inside a black hole, the singularity is inevitable in classical general relativity. It remains an open question how this singularity can be resolved or understood in a complete theory which can describe gravity and quantum mechanics in a consistent way (quantum gravity).

2.4 Gravitational Waves

Gravitational waves are consequences of the full theory of general relativity. Deriving them rigorously is beyond the scope here. We nevertheless build some intuitive understandings.

Special relativity and Newtonian gravity are inconsistent

What's wrong with the good old Newtonian gravity $F = G_N M m / r^2$?

Suppose Alice is a light year away from Bob and she waves her hand. As the position r of her hand changes, Bob can in principle measure a slightly different gravitational attraction from Alice *immediately*—because force needs no time to propagate in Newtonian

---
[①] This explains item ⑤ at the beginning of this part.

gravity. Alice can thus encode information in how her hands move, and the information is sent to Bob immediately, which is faster than light.

From special relativity, no information can travel faster than light. Thus, Newtonian gravity is not consistent with special relativity.

Conceptually, replacing gravity by spacetime curvature does not solve this problem — can the spacetime curvature change with a speed faster than light when you move your hands?

Early history of gravitational waves

Back to 1893, Oliver Heaviside noted the similarity of gravity and E&M, and suggested that gravity may propagate in waves. But only after the establishment of general relativity, the discussion of gravitational waves was put into the appropriate scientific framework. Einstein was the first to predict gravitational waves in 1916 based on his general relativity.

What can gravity learn from E&M?

In Newtonian gravity, you can send superluminal information by waving your hands and detect the force change using $F = G_N M m / r^2$. However, why don't E&M have the same problem with Coulomb force $F \propto Qq/r^2$?

If we only have the Coulomb force formula and nothing else, we would have the same superluminal problem for E&M. However, the magnetic field comes to rescue here. Once the source electric charge accelerates (you cannot encode information by inertial motion without acceleration), the time variation of the electric field produces magnetic field, and the time variation of the magnetic field produces electric field. This process happens over and over again and E&M wave is emitted at exactly the speed of light. Also, in this way, the E&M field can propagate by itself without the need of matter (after produced by a source) as new "degrees of freedom".

The lesson: we hope that new "degrees of freedom" can come to rescue in gravity, similar to what magnetic field does for E&M. Where to find those degrees of freedom?

Are gravitational waves physical?

The concern is that space deforms, so does rulers. So can we actually measure gravitational waves? The debate lasted for 40 years. In 1957, Feynamn (under a pseudonym of "Mr.

Smith") argued that when gravitational waves pass, a waterdrop on a rod will move wrt the rod and thus generate heat by friction. Thus gravitational waves are physical. The observational efforts start shortly afterwards. In 2016, after decades' trial and failure, gravitatioanl waves are finally detected by the Laser Interferometer Gravitational-Wave Observatory (LIGO) using an interferometer similar to the Michelson-Morley experiment.

(Optional) The dynamical metric and the gravitational waves

We have shown that a static gravitational field can be described by a non-trivial metric, for example Eq. (2.5). There are other components in the metric as well. The metric can be viewed as a 4×4 symmetric matrix with 10 components (free functions to choose). 8 of them are already used:

(1) 4 to describe the response to energy and momentum of matter. The energy part is the Newtonian gravity and the momentum part is its relativistic generalization. Technically they are called the Hamiltonian (energy) and momentum constraints.

(2) 4 represents the freedom to choose coordinates. When choosing any new coordinate system $\tilde{x}^\mu(x)$, we expect that the physical interval should not change and thus the metric should transform accordingly.

What are the remaining 2 degrees of freedom? Happily they correspond to propagating waves of gravity, to solve the problem of Newtonian gravity. And the gravitational waves propagate at the speed of light.

What do gravitational waves look like?

As we have argued, the gravitational waves are the fluctuations of spacetime, characterized by the metric. Thus, their impacts should be the distortion of distances. They travel at the speed of light and they are transverse waves (polarized perpendicular to the propagation direction). The two independent polarization modes deforms spacetime in Fig. 2.3.

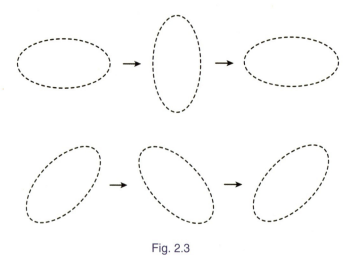

Fig. 2.3

2.5 Epilogue: Summary and What's Next

A rare example in science

General relativity is a very rare example in the history of modern science, that a completely new scientific framework is setup (almost) purely by logical reasoning and philosophical belief.

Starting from the equivalence principle, we have shown you how time becomes relative in a uniform gravitational field. Then we argue that there is a similar effect for spherical stars. And if the star is dense enough, black holes form. The black hole has an event horizon where matter (classically, without the quantum mechanical considerations) can enter but not return. From the similarity between gravity and E&M, we have argued that gravitational waves exist as deformations of spacetime. All these can be put into precise and elegant forms with the mathematical framework of differential geometry.

The equivalence principle is the first observation in general relativity, but not yet the punch line. Though beyond the scope of the course, let's mention the two key concepts in general relativity:

How space curves?

How is a geometry such as Eq. (2.5) determined? In Newtonian gravity, we know that gravity is determined by matter distribution. Now that gravity is interpreted as geometry, geometry should be determined by matter distribution as well, at least in the Newtonian limit. The rule to determine geometry from matter distribution is the Einstein's equation. It looks like $G_{\mu\nu} = 8\pi G T_{\mu\nu}$ and is the master equation of general relativity. I cannot explain it to you within the scope of this course, except mentioning that the LHS is determined by the metric $g_{\mu\nu}$ and its derivatives; and the RHS is determined by matter distribution. This is dubbed "matter tells space how to curve".

How matter moves?

How matter moves in curved space? Matter follows the geometry of spacetime. This is dubbed "space tells matter how to move". Given a spacetime geometry, the motion of free-falling matter is determined by a generalized version of the twin paradox. Recall that in the twin paradox, the free observer without external force exerted has the longest proper time. The equivalence principle implies that, free-fall observers in general relativity still have extremal (usually longest but there are exceptions that the proper time is only longest for a selection of variations) proper times. Such extremal paths are known as geodesics. A geodesic can be interpreted in a semi-Newtonian manner as follows: the equation of a geodesic is $a^\mu = -\sum_{\alpha,\beta=0}^{3} \Gamma^\mu_{\alpha\beta} u^\alpha u^\beta$. Here a_μ is the 4-acceleration and u^α is the 4-velocity. $\Gamma^\mu_{\alpha\beta}$ is a generalization of the concept of gravitational force, made from the metric and its derivatives. You may have recognized that this looks like the Newton's second law. It is at your well whether to interpret the RHS as a gravitational force, or a term in geometry to tell matter how to move. They are equivalent as implied by the equivalence principle.

Is general relativity general?

Greneral relativity applies to all known scales where gravity can be studied, including millimeter scales for very precise gravity experiments in the lab, to the size of the whole universe.

Having that said, so far, we still do not understand how general relativity can be built within the framework of quantum mechanics. The efforts in searching for such a unified theory is known as "quantum gravity". Experiments have not been precise enough to probe quantum effects of gravity. But theoretical consistence between general relativity and quantum mechanics is also a well motived question, and is surprisingly hard. The current leading candidate of quantum gravity is string theory, which begins with the hypothesis that elementary particles are actually strings with one spatial dimension instead of zero (zero corresponds to point particle).

Further reading

There are many great books of general relativity. Some simpler ones for beginners include Gravity: *An Introduction to Einstein's General Relativity* by Hartle and *A First Course in General Relativity* by Schutz.

Exercises

E2.1

Calculate how much faster your upstairs grow older than you. Take into account both special relativity and general relativity effects.

E2.2

Given the precision of modern satellite navigation systems (such as BeiDou, Galileo, GLONASS or GPS), how much does special relativity and general relativity effect affect the precision of these satellite navigation systems, if these factors were not taken into account?

Chapter 3
Cosmology

The comprehensible universe

The universe is the most complicated object in our universe, since it contains all complexities of all objects in our universe. However, amazingly, Einstein says:

"The most incomprehensible thing about the world is that it is comprehensible."

Here let us try to understand the following questions:

① Before understanding everything in the universe, how can we understand anything about the universe, since it is the most complicated?

② Why is the night sky dark?

This question looks stupid, as we are not under sunshine at night. However, what about stars? Why doesn't the night sky appear as bright as daytime under the shining of the stars?

You may answer: they are far away. Thus their apparent luminosity decays as $1/r^2$. So they are not as bright as the sun.

Correct. However, how many stars are there at distance r? If our universe is homogeneous and infinite in space and time, statistically, the number of stars at distance r should scale as r^2. So why doesn't their light sum up and being as bright as the sun?

This question is known as the Olbers paradox (Digges, 1576; Olbers, 1823).

> **Other ways to Olbers paradox**
>
> The Olbers paradox can also be understood in other equivalent ways, less obvious but may be more insightful:
>
> Within an infinitesimal solid angle cone along your line of sight, a star is as bright as the sun. The sun is overall brighter because it spans more solid angle. If the universe is infinite, sooner or later the cone along your line of sight will reach a surface of a star. Thus you will see the sky everywhere as bright as the sun.
>
> Yet another way to the Olbers paradox: if the universe is static and has infinite duration, we will reach equilibrium with the stars thus the earth is as hot as the star surface.
>
>

3.1 The Dynamics of the Universe

The cosmological principle

To make the universe simpler, let's assume:
Our universe is approximately homogeneous and isotropic on large scales.
This is known as the cosmological principle.

The cosmological principle had already appeared in Newton's Principia (1687). However, the study of cosmology in Newtonian mechanics is full of paradoxes. The framework of modern cosmology is setup based on Einstein's general relativity.

In this section, we will still use Newtonian mechanics to set up the theory of our universe with special care. The subtleties in the discussion can be resolved in general relativity.

Distant galaxies are leaving us

In the 1920s, Hubble (see also Lemaitre and other astronomers' contribution) measured the spectrum of light from other galaxies. As we know the spectrum at emission (recall the characteristic spectrum by atoms), by measuring the observed spectrum, one can find out the velocity of these galaxies along the line of sight direction via Doppler-like effects.

The result shows that the distant galaxies are leaving us.

> **Faster than light?**
>
> Is the expansion of the universe faster than light? To be more precise, for distant enough object can they leave us at a rate faster than light?
>
> Yes or no. That depends on how to define speed in general relativity.
>
> If you define speed as $dR(t)/dt$, then for distant enough objects, they are indeed leaving us at a speed faster than light. This is not due to the objects themselves moving fast, but instead the emerging space between objects. This is not contradictory with special relativity, since the speed between two near-by objects (where special relativity applies) cannot be faster than light.
>
> If you define speed as the speed of the moving object (peculiar velocity), excluding the expansion of space, then the speed of distant objects are still bounded by the speed of light.
>
>

From the Copernicus principle to the expanding universe

Since Copernicus, human start to realize that the universe does not have a center. However, if all the distant galaxies are moving away from us, does that mean we are at the center of the universe?

Not necessarily. Imagine an expanding membrane, or balloon. Wherever you are on the membrane or balloon, you find your neighbor points leaving you. Thus, we can interpret the observation as that the universe is expanding.

Now that the universe is expanding, how is the expansion rate determined?

Let's imagine a universe filled with pressureless dust. The dust particles expand together with the universe expansion, without additional motion by themselves (i.e. comoving, i.e. without peculiar motion, Fig. 3.1).

To quantify the expansion rate of the universe in the framework of Newtonian mechanics, define two set of distances: comoving distance and physical distance.

(1) The physical distance $R(t)$ is the real distance between two objects, corresponding to actual physical measurements.

(2) The comoving distance r is measured by an expanding coordinate system, such that the comoving dust particles have fixed (i.e. time-independent) coordinate value in

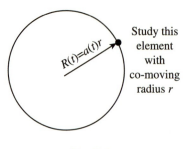

Fig. 3.1

the comoving coordinates.

The comoving coordinate is related to the physical coordinate by $R(t) = a(t)r$, where $a(t)$ is known as the *scale factor*, quantifying the time dependence of the universe. The expansion rate of the universe is quantified by the Hubble parameter:

$$H(t) \equiv \frac{\dot{a}}{a} \tag{3.1}$$

The universe tells matter how to get diluted

Consider a component in the universe (for example, radiation or dust particles) with energy density ρ and pressure p. How does ρ evolve with time?

The first law of thermodynamics is $dE = TdS - pdV$. The expansion of the universe is adiabatic since heat will not be conducted in a homogeneous universe. Thus $dS = 0$. We have, during the expansion of the universe, $dE/dt = pdV/dt$. Here V is a physical volume $V \propto a^3$. Thus,

$$\frac{d(a^3 \rho)}{dt} = p\frac{da^3}{dt} \quad \Rightarrow \quad \dot{\rho} + 3H(\rho + p) = 0 \tag{3.2}$$

This is known as the continuity equation in cosmology, telling that how matter gets diluted during the cosmic expansion.

> **Dust and radiation**

As examples, we note that

Pressureless dust: $p = 0$. Solving (3.2), we get $\rho \propto a^{-3}$. The particle density is diluted as a^{-3}, and the energy per particle is a constant.

Radiation: The photon density $\propto a^{-3}$. However, the wavelength of each photon is stretched as $\lambda \propto a$. Thus the energy per photon is proportional to $1/a$ (quantization condition). As a result, $\rho \propto a^{-4}$. This corresponds to $p = \rho/3$.

As radiation gets diluted faster than dust, our universe transited from radiation domination to dust domination during its expansion.

What's the matter in the universe?

We know only 5% of our universe (weighted by their energy composition). Fig. 3.2 is a figure indicating what we know and what we don't know.

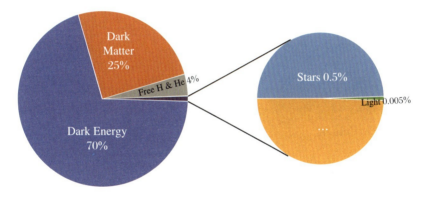

Fig. 3.2

(1) The matter that we know: Matter made by nucleus and electrons (either in the phase of atoms or plasmas) takes 5% of the energy budget of our universe. Most of these atoms and plasmas are in the form of free Hydrogen and free Helium. A small part forms stars. Among the matter that we know, the weight of light also takes a tinny part (0.005%).

(2) The matter that we don't know: observations indicate that there are dark matter (takes about 25%) and dark energy (takes about 70%) in the universe. We don't know what are they. But from observations, we know approximately their nature: the dominate part of dark matter is almost pressureless with $p \sim 0$, similar to that of atomic matter now, and contribute attractive gravitational force. However, dark energy has negative pressure $p \sim -\rho$, which effectively provide repulsive gravitational force and is driving an accelerated expansion phase of our universe.

Einstein's "biggest mistake"

When Einstein first used general relativity to study our universe, he found that the universe will either expand or contract. He did not like the situation. To keep the universe static as he likes, he added a cosmological constant to his equations (1917). After Hubble discovered the expansion of the universe in the 1920s, Einstein regretted that he had missed the chance to predict an expanding universe, and considered the introduction of the cosmological constant the biggest mistake in his life (1931). However, in 1998, dark energy is discovered, and by far the simplest candidate is the cosmological constant (with a different value, which does not make the universe static but make it expanding with positive acceleration).

Matter tells the universe how to expand

To determine the dynamics of $a(t)$, we consider a dust particle with mass m, located at comoving distance r (and thus physical distance $R(t) = a(t)r$). The energy conservation equation for this particle is

$$\frac{m}{2}\left(\frac{\mathrm{d}R}{\mathrm{d}t}\right)^2 - \frac{GMm}{R} = E = \text{constant} \tag{3.3}$$

The constant E is determined by the initial condition of the universe. Observations tell that $E \approx 0$ thus we will ignore it (but do not ignore the question why $E \approx 0$).

Here M is the mass inside the sphere:

$$Mc^2 = \frac{4\pi R^3}{3}\rho \tag{3.4}$$

What about the mass outside?

One can argue that the mass outside the sphere has no force on our particle m, since the force cancel for each spherical cell. In fact, the statement depends on the boundary condition of the universe. This is the limitation of the Newtonian cosmology. One needs general relativity to remove this boundary condition dependence.

Inserting Eq. (3.4) into Eq. (3.3), we get the Friedmann equation telling the universe how to expand:

$$H^2 = \frac{8\pi G\rho}{3c^2} \qquad (3.5)$$

In Newtonian cosmology, we cannot deal with $p \neq 0$. This is because Newtonian gravity does not define how pressure contributes to gravity. Thus in the above derivation, we used a dust-filled universe. In general relativity, one can verify that (3.5) actually also holds for $p \neq 0$, for example a radiation filled universe.

As solutions of Eq. (3.5), for a dust-dominated universe, $a(t) \propto t^{2/3}$ and for a radiation-dominated universe, $a(t) \propto t^{1/2}$. We will leave the details to an exercise.

The age of the universe

We observe that for dust and radiation dominated universes, at $t \to 0$, the scale factor approaches zero and thus the energy density diverges. Thus, we consider $t = 0$ as the start of the universe. More detailed study of the very early universe changes this starting point a bit but not very much.

For a matter dominated universe, from the definition of the Hubble parameter, we have $t = 2/(3H)$. To include the observed dark energy (a component with $p = -\rho$), the age of the universe is corrected to approximately $t \approx 1/H$. Thus, the age of the universe is closely related to the Hubble parameter — the expansion rate of the universe.[1]

3.2 The Early Universe

When you are looking deep into the night sky, you are looking through the history, since the star lights may take years, or tens of thousands of years to travel before reaching us. Modern telescopes can extend this record greatly, to billions of years. What does the early universe look like?

---

[1] For recent estimates of the universe, in 2012 the WMAP experiment estimated that the universe is 13.77 billion years old; and in 2018, the Planck experiment estimated that the universe is 13.8 billion years old.

Thermal history: the earlier, the hotter

The earlier universe is hotter than our current universe. That's because the expansion of the universe stretches the energy per photon. Or you can understand it as that during the expansion the gas does work and thus lose energy.

The events in the early universe is related to the temperature scales of the universe. Thus, the history of the universe is known as the thermal history.

It is believed that the very early universe was in a state with a higher temperature than all man-made experiments now and in the forseeable future.

Fig. 3.3 is about the thermal history of the universe.

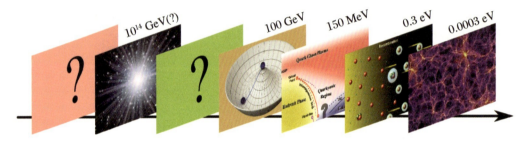

Fig. 3.3

(1) When the universe was 20 picoseconds old (with temperature 100 GeV), a spontaneous symmetry breaking generated mass for known massive fundamental particles.

(2) When the universe was 20 microseconds old (with temperature 150 MeV), free quarks are binded into protons and neutrons. Their binding enery is the main source of mass of atomic matter.

(3) When the universe was 3 minutes old (with temperature 0.1 MeV), light elements, especially helium, are created from protons and neutrons.

(4) When the universe was 50,000 years old (with temperature 1 eV), the universe becomes dominated by non-relativistic matter instead of relativistic radiation.

(5) When the universe was 400,000 years old (with temperature 0.3 eV), the universe become transparent and light for the first time can travel in the universe freely.

(6) In the following billions of years, structures grow in the universe.

(7) Now our universe is about 14 billion years old. The universe starts to be dominated by dark energy with $p = -\rho$. We yet need to theoretically and observationally understand the nature of dark energy.

> **Is energy conserved in cosmology?**
>
> This is a very tricky question. We can answer it in a few aspects:
>
> In cosmology, there is no time translation symmetry, since $a(t)$ depends on time. Thus energy will not arise as a conserved quantity following the Noether theorem.
>
> We still have approximate local energy conservation if restricted to very small space volume and time duration: $\sum_\mu \nabla_\mu T^{\mu\nu} = 0$.
>
> The matter energy drops since matter has positive pressure and does work when the universe is expanding.
>
> If the gravitational potential energy is considered together, the answer becomes indefinite since there are multiple ways to define gravitational energy in general relativity.
>
>

3.3 Epilogue: Summary and What's Next

Further reading

See Baumann's *Cosmology* for a detailed introduction of this content.

What happens next?

Let's first literally discuss what happens next — what's the future (even fate) of the universe? Now our universe is dominated by dark energy. The energy density of dark energy does not change during the expansion of the universe. As a result, dark energy will become more and more important for the fate of the universe. If the energy density of dark energy is indeed a constant, most galaxies in our present universe will leave our horizon, leaving only $\mathcal{O}(100)$ galaxies around us observable (known as the local group). But we need to further understand dark energy both theoretically and observationally to be more confident about this fate.

You will find more details about our universe in a course of cosmology. Also, in general relativity there is usually an introduction to cosmology as an application. In-depth study of cosmology usually has close relations to astronomy and high energy physics.

Exercises

E3.1 Scale factor as a function of time

For $p = 0$ (dust), $p = \rho/3$ (radiation), $p = -\rho$ (dark energy), solve the Friedmann equation to get $a(t)$. Based on your solution, estimate the age of the universe given the value of H.

E3.2 The age of our universe

For the energy composition given in the pie diagram of the main text, estimate the age of the universe. You may first have an analytical estimate and then numerically solve the equations to test your estimate.

Chapter 4
Quantum Mechanics

Alice's adventures in a quantum wonderland

Alice drinks a bottle of "drink me" and becomes as small as an atom.[①] She finds the world totally unfamiliar, compared to the familiar classical world when she was of normal size.

① When she knows where she is, she doesn't know how fast she is walking; When she knows how fast she is walking, she doesn't know where she is.

② Walking in the brightness helps her to know better where she is.

③ Walking in the brightness, Alice finds the apparently continuous beams of light hits her as bullets: one shot after another in a discrete way. And she feels hurt more by the "bullet" of blue light than the "bullet" of red light.

④ She no longer has to enter a room through a door. She has a small chance to cross the wall and enter the room directly (although more likely she gets bounced back).

⑤ With fine-tuned speed, she can greatly enlarge the wall-crossing chance by using the front and back wall of the room together.

⑥ If she still prefers to walk through doors, she can walk through two doors at the same time. She is then likely to appear in the room somewhere, but never somewhere else.

⑦ She met an electron and made friend with him. However, soon, Alice was unable to find this electron friend out from other electrons no matter how hard she tries.

[①] I don't know how to talk about a person as small as an atom consistently, but let us imagine that she is now just an atom, but somehow she can still record and tell her experiences.

In this part, let us find out what happens in the microscopic world, where the laws of nature is quantum mechanics. Let me tell that the learning experience of quantum mechanics will be very different from special relativity. Knowing it in advance helps for your learning.

Quantum mechanics and special relativity are of different "feeling"

(1) Special relativity is based on 2~3 simple postulates — everyone knowing general physics knows what they mean physically. The logical sequence of these postulates may be counter intuitive. But after thinking, you can know what these consequences mean physically and that provides you a unique way to model them explicitly in your mind.

(2) Quantum mechanics is based on about 5 fundamental postulates.[1] Each postulate looks mathematical. You may scroll to the summary of this part to get a quick feeling, not to be scared by their appearances. They tell you how to compute things. But they do not give you a physical idea in your mind (known as interpretations of quantum mechanics) to think about what's really happening.[2]

Due to the differences, here we will not start by throwing the postulates to your face. Instead, we will spend longer to explain why we are forced to impose these "exotic" postulates, by reviewing key experiments and their explanations.

4.1 The Nature of Light

4.1.1 Is Light Particles or Waves?

The nature of light was studied and debated for thousands of years. In the classical era (until 1900s), the key debate about light can be summarized as: Is light a particle or a wave? To be clear, let us summarize what particles and waves are. I encourage you to tap a basin of water or throw stones into a lake, to watch the nature of classical waves if

---

[1] The exact number of postulates of quantum mechanics depends how you formulate it. You will see 4, 5, 6 in different books.

[2] The interpretations of quantum mechanics is still an open question. There are a few possible interpretations in the market now. But we don't know which is correct (and we don't even know if it's possible to figure out the correct one).

you haven't done so before.

> **Nature of classical particles**
>
> Can be counted and labeled as $1,2,3,...$.
>
> Have definite locations (or mass centers).
>
> Have energy & momentum with $E = p^2/(2m) + V$.
>
> When two particles meet, they collide.

> **Nature of classical waves**
>
> Is oscillation with continuously variable amplitude.
>
> Extended objects without definite locations.
>
> Have frequencies ν and wavelength λ.
>
> When two waves meet, they overlap on top of each other.

Now that you know very well what classical particles and waves mean, let us explore the nature of light.

Is light particles or waves? The debate in the history

(1) As early as 400BC, Democritus asserted that all things are made of atoms (tinny particles) including light.

(2) In the early 17th century, optics was developing very fast, led by microscopes, telescopes and so on. In 1630s, Descartes considered light as wave. In 1660, Grimaldi discovered diffraction of light, which behaves like, water wave.

(3) Newton discovered the dispersion of light (1666). He considered light as particles and dispersion is interpreted as mixing and separation of different particles (1672).

(4) Hooke and company strongly criticized Newton's particle theory of light and supported the wave theory.

(5) In 1704, Newton published "Opticks" (Hooke died in 1703). Due to his impact at that time, the particle theory of light dominated.

(6) In 1801, Young made a double-slit experiment. After that and further developements (for example the Poisson spot) the wave theory become dominate again.

(7) In 1861, Maxwell published his equations of E&M. Electric field, magenetic field and light are then unified in one theory, where light appeared in the form of wave solutions.

Maxwell's equations conclude the debate about the *classical* nature of light — light is wave, and its mathematical equations are found from the first principle. However, the nature of light starts to look surprisingly different again once we step into the modern era.

The whole thing started at an "ultraviolet catastrophe". The observed radiation from black body does not agree with Boltzmann's theory of statistical physics for short wavelength light. In 1900, Planck suggested that light is emitted and absorbed in a quantized way to solve the problem. This is the first indication of the quantum world historically. Here we will not follow a historical order, but instead show you two more intuitive experiments made slightly later, where classical physics fails.

The double-slit experiment

The double-slit experiment is an important turning point in the history for the wave theory of light. Let us review it here. As in Fig. 4.1, two beams of light travel through two slits and meets on the screen. The experiment needs to be explained by the wave theory of light.

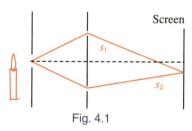

Fig. 4.1

On the screen, the pattern looks is shown in the following picture.

The position of the bright and dark bands on the screen can be computed by comparing distances s_1 and s_2 on the figure.

If $s_1 - s_2 = 0, \lambda, 2\lambda, \ldots$, then light from both trajectories oscillate towards the same direction (constructive interference), and they leave bright bands on the screen.

If $s_1 - s_2 = \lambda/2, 3\lambda/2, \ldots$, then light from both trajectories oscillate towards different directions (destructive interference), and they leave dark bands on the screen.

For the double-slit experiment to work, the light wave must at the same time go through two slits. If we block one slit, the originally dark spots will become bright again.

4.1.2 The Photoelectric Effect

The photoelectric effect

Consider the experiment in Fig. 4.2. A beam of light shines on metal. Apart from reflection light, what else do we expect to see?

(1) In 1887, Hertz noted that electrons can be knocked out by light. Such electrons are called photoelectrons. He also noted that the effect needs ultraviolet (UV) light.

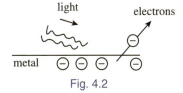

Fig. 4.2

(2) In 1902, Leonard studied the effect in more details. The energy of individual photoelectron increases with the frequency of light, but independent of the intensity of light. For each type of material, there is a cutoff frequency ν_0, below which there is no photoelectron emission at all. The situation is plotted in Fig. 4.3.

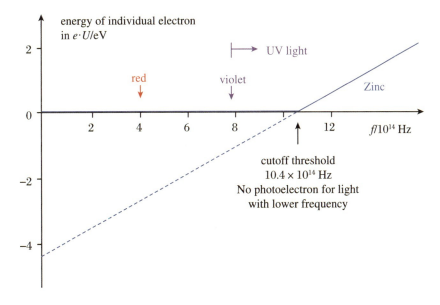

Fig. 4.3

> **Summary of photoelectric effect**

Dependent on	I	ν
Number of photoelectron	Y	N
Energy of photoelectron	N	Y
threshold	N	Y

Here I and ν are the the intensity and frequency of the incident light, respectively.

Why can UV light with high enough frequency knock out electrons? While light with lower frequency cannot knock out electrons no matter how strong the beam of light is. This surprising effect cannot be explained in classical E&M, and gives us a clue about quantum mechanics.

Honestly, when I learned this I did not feel surprised though the textbook told me to feel so. Trust me, it's indeed surprising. If you are not convinced, let's instead think about the following game.

The surprising whac-a-mole

Whac-a-mole is a game to knock mice back to their holes. As physicists, instead of thinking how to knock them back, let's model why they come out. Assume they come out because they are scared by earthquake (Fig. 4.4).

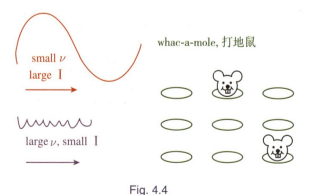

Fig. 4.4

Imagine two types of earthquakes:
(1) Red earthquake has low frequency but very very large shaking amplitude.
(2) Violet earthquake has high frequency but very very small shaking amplitude.
Which earthquake is more likely to scare mice out?

In reality, the red earthquake is of course more scary. But if we correspond this game to the photoelectric effect experiment, what the experiment tells is that

(1) The violet earthquake can scare some mice out despite of the small amplitude it has (indeed if it is stronger it can scare more mice out).

(2) No matter how strong the red earthquake is, it does no scare *any* of the mice out.

Intensity still matters

When photoelectrons can come out, the number of photoelectrons indeed depend on the intensity of light. The stronger light, the more photoelectrons. But the key puzzle is why the threshold of whether photoelectrons come out does not depend on intensity.

The photoelectric effect cannot be explained by classical E&M

Now have you started to feel that the photoelectric effect is surprising? Classical E&M cannot explain photoelectric effect. In classical E&M, electrons are shook by light, just as the mice shook by earthquake. When the intensity (i.e. shake amplitude) of the light is large enough, some electrons are shook so hard that they fly out of the metal. However, now whether electron leaves the metal depends on frequency instead of intensity. This poses a serious problem for the classical theory of light.

Quantities describing waves

Wavelength λ: length of a period of oscillation.

Wavenumber $k \equiv 2\pi/\lambda$: number of radians the wave oscillates per unit distance.

Period $T \equiv \lambda/v$: time duration of a period of oscillation. Here v is the phase velocity of wave. For light, $v = c$.

Frequency $\nu \equiv 1/T$: number of oscillations per unit time.

Angular frequency $\omega \equiv 2\pi\nu$: number of radians the wave oscillates per unit time.

Einstein's explanation of the photoelectric effect

Inspired by Planck's 1900 explanation of black body radiation, Einstein proposed in 1905 that in the photoelectric effect, light can be considered as a collection of tiny particles (now known as *photons*). Each photon has energy

$$E = h\nu = h\frac{c}{\lambda} = \hbar\omega \tag{4.1}$$

where $h = 6.6 \times 10^{-34}$ m$^2 \cdot$ kg$\cdot$s^{-1} is the Planck constant, and $\hbar = h/(2\pi) = 1.1 \times 10^{-34}$ m$^2 \cdot$ kg$\cdot$s^{-1} is the reduced Planck constant.

Eq.(4.1) explains the photoelectric effect (and Alice's experience ③ in her wonderland). Because an electron is likely to be hit by one photon instead of multiple photons together (if a person is hit by car, it is almost impossible that the person is hit by two cars at the same time). Thus whether the electron is knocked out of the metal is determined by the energy of the photon E, and thus its frequency ν, instead of the intensity of light.

> **Photon, quanta and quantization**
>
> The "tiny particle" as building block of light is called photon.
>
> In general, the "tiny particle" (or quasi-particles where there is no actual particle) as building block of some form of energy (including matter itself, or the rotation, oscillation of matter, etc) is called quanta. In this sense, photon is the quanta of light. We will see that there are many other types of quanta that makes up our world.
>
> The feature that the quanta can be counted one by one (instead of being continuous) is that the quanta are "quantized".

"Classical" light can be considered a "coherent state" of many photons. In a rough sense, coherent means that the E and B fields of the photons add coherently (instead of cancel each other in a disordered way).

The everyday photoelectric effect

We actually don't have to do experiments to find out the photoelectric effect. We know that strong sunlight hurts our skin. To reduce the hurt, we can use sunscreen (sun cream). For example, after applying SPF 30 sunscreen and stay under the sun for 30 minutes, in principle, your skin damage equals to staying 1 minute under the sun.

How SPF 30 sunscreen works? To prevent damage to your skin (prevent sunlight reaching your skin), it should do either of the below for you:

Reflection: reflect 29/30 of the light away. But, why don't you look like a mirror after applying sunscreen?

Absorption: absorb 29/30 of the light before it reaches your skin. But, why don't you look black after applying sunscreen?

This is because sunscreen does not reflect or absorb all frequencies of light. It is much more effective on UV light. Indeed you look different after applying sunscreen if you take a photo in the UV light frequency band.

Why does the UV light hurt your skin more than the visible light (which dominates energy of sunlight)? The reason is the same as the photoelectric effect (but what get knocked is not electrons but chemical bonds, Fig. 4.5).

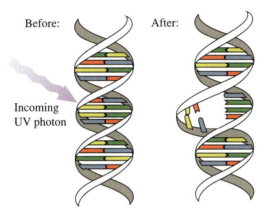

Fig. 4.5

Also, in the winter, when you get close to a heater, you feel as warm as being under the sun. However, you don't have to apply sunscreen around a heater. This is because the heater radiation is dominated by the IR light.[①]

Does the photon wavelength change? The Compton effect

In the photoelectric effect, we have discussed how the electrons behave when light shines on them. What about the photons themselves? Let the incoming photon wavelength be λ and the outgoing photon wavelength be λ' (Fig. 4.6). Are they equal?

In classical E&M, $\lambda = \lambda'$. The incoming light (which has oscillatory electric field) shakes the electrons, then the shaken (i.e. accelerated) electrons runs away, and at the

---

① And also you do not have to apply sunscreen when using your phone or microwave ovens, either.

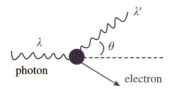

Fig. 4.6

same time outgoing light of the same frequency emits.

When the electrons got hit by light, they get additional kinetic energy. Where does the additional kinetic energy come from? In classical E&M, the kinetic energy come from the reduction of the intensity of light, i.e. the outgoing light has small intensity.

However, the above energy argument must be wrong in a quantum theory. A photon cannot further reduce its intensity. Then, where would the kinetic energy of the electron come from? A photon has nothing to lose but its frequency (recall $E = hc/\lambda$). Thus we must have $\lambda' > \lambda$. As a good exercise, assuming that the electrons are free without any binding energy, then energy-momentum conservation together with quantization $E = hc/\lambda$ gives

$$\lambda' - \lambda = \frac{h}{m_e c}(1 - \cos \theta) \qquad (4.2)$$

Up to the angular factor $1 - \cos \theta$, the shift of wavelength is $\lambda_C \equiv h/(m_e c) \approx 2.4 \times 10^{-12}$ m, known as the Compton wavelength of an electron. The effect is observed by Compton (1923) and known as the Compton effect. Since the Compton wavelength corresponds to X-ray wavelength of light, the effect is hard to observe for observable light, and clearly visible when the wavelength of light is as short as X-ray.

The dimension of $\hbar$

It is interesting to note the dimension of $\hbar$: it connects mass, speed and length. This relates to the question: why there is a fixed size of atom, such that the electron does not fall into the nuclei. From dimension analysis, the electron mass is a fundamental quantity of nature. Through speed of light c (enter from Coulomb's law) and $\hbar$ (and a dimensionless strength of interaction α), the electron mass scale m_e got converted to the atomic length scale $\hbar/(m_e c \alpha)$, known as the Bohr radius. Thus, from dimension analysis, before putting in any dynamics, we already except that we need such a fundamental constant. We will see it explicitly in the part of atoms.

> **Planck scale and quantum gravity**
>
> In addition to the Bohr radius, we can also make another scale out of $\hbar$ once gravity is concerned. The scale is the Planck energy $E_\mathrm{P} = \sqrt{\hbar c^5/G} \sim 10^9$ J ~ 500 kW·h, approximately the kinetic energy of an operating train. This scale puts together quantum and gravity, and is the scale of quantum gravity.
>
> Consider two electrons, each has kinetic energy comparable to a running train, collide head-to-head. The collision is energetic enough to produce an microscopic black hole. Such collisions need to be studied with quantum gravity. We still don't have a full understanding of quantum gravity, though there is a decent candidate known as string theory.
>
>

Wait! Haven't you said that the particle theory of light is dead?

In the last subsection 4.1.1, I have already told you that the Newtonian particle theory of light is dead. Now, why does Einstein dare to use particles to explain light again?

The photon that Einstein proposed is not a "classical particle" as we usually imagine. It has the feature of particles in that its energy is quantized and it appears to interact with the electron in the same way as two point particles collide.

Haven't I said that the particle theory of light is killed by the double-slit experiment? Now let's see how the photons behave in the double-slit experiment.

4.1.3 The Single Photon Double-Slit Experiment

The single photon double-slit experiment

In section 4.1.1, we have reviewed the double-slit experiment. Image that we reduce the intensity of light. Eventually, in the extremely-low intensity limit, each time we are doing double-slit experiment with only one photon. What is the outcome of such a single photon double-slit experiment? Suppose that we put a sensitive film on the screen which can record every photon. What's the status of the film as time goes on?

The experiment was carried out in 1909 by Taylor. The result looks like the figures in Fig. 4.7. Two features are clearly spotted in the result:

(1) At the early stage of the experiment (panels (a) and (b) of the figure), clearly, the photon postulate is supported: Indeed we see a few "particles" on the screen. Each photon leaves one point on the screen.

Fig. 4.7 Single quantum double-slit experiment. Here electrons are used. Photons behave similarly.

(2) As the experiment goes on (from panels (c) to (e)), more and more photons pass the slits. Interference patterns emerges. Indeed there are bright bands where the photons are more likely to reach and dark bands where the photons cannot reach.

How to interpret this result?

Considering that the intensity of incoming light is low enough, one photon already reaches the screen and leaves an unchangeable record before the other photon starts off. Thus the photons can be considered as independent.

Each photon behaves independently but ends up at different points on the screen for the same experimental setup. Thus there must be some randomness in the photon behavior.

So the position of a photon on the screen should satisfy a probability distribution (as a smoothed version of panel (e)). The probability is large at bright bands and vanishes at dark bands, forming interference patterns.

You may ask where the randomness comes from. We will come back to this point when discussing quantum measurement. The short answer at the moment is that there is no unique fundamental answer to the origin of randomness. But we have a well-defined way to compute the probability distribution. In this section, let us focus on another question, along the line of particle-wave debate of light.

The which-way experiment

You may want to further see which slit on earth the photon goes through. To do so, you put a detector on one of the slit. If the photon goes through this slit, the detector reports it. However, once a detector is placed, the interference pattern vanishes.

There are many more confusing results, including compromise which-way experiment (1987), delayed choice (1999), weak measurement (2012), and so on. But here we have enough surprises to explore at the moment and will not discuss these experiments here.

Is photon a particle or a wave?

If you think the photon is a particle in the classical sense. Then recall how the interference patterns are formed:

The photon must have gone through both slits. But how can a particle go through both slits at a time? Note that the photon is not supposed to further split into even smaller particles. Otherwise that contradicts the photoelectric effect, and contradicts the fact here that each photon leaves one point on the screen.

The dark band of the interference pattern indicates destructive interference (cancellation between two branches). When particles meet, they collide. How can they stay at the same position and cancel each other?

So you probably change your mind and consider photon as a wave in the classical sense. Then

Why do photons arrive at the screen one by one, and one photon is the smallest building block of energy? Can't we continuously reduce the energy of wave by continuously reduce the amplitude of oscillation?

Why does each photon leave one point on the screen, instead of a weak but complete interference pattern?

(Optional) From double-slit to path integral

One can generalize the single-photon double-slit experiment by adding more slits on a board A (Fig. 4.8)

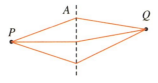

Fig. 4.8

more boards $A_1, A_2, \ldots, A_n$ (Fig. 4.9)

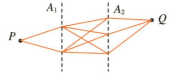

Fig. 4.9

The interference pattern at Q is the phase interference among all these paths. This should still apply if we have an infinite number of boards and each board has an infinite number of slits on it.

What does a board with an infinite number of slits mean? It can be considered as free space with no board at all! The probability amplitude of photon travelling in free space between P and Q can be computed by summing over the interferences of all paths. This approach is known as the path integral formulation of quantum mechanics, as to be discussed in the part of action principle.

4.1.4 A Wave-Particle Duality, and from Light to All Matter

A wave-particle duality

Is a photon a particle or a wave? This is a wrong question to ask.

We have listed the properties of particles and waves in section 4.1.1. But who told us that microscopic matter must be classified into one of them following our classical experience?

It turns out that microscopic matter has some natures of classical particles, and some natures of classical waves together. The full properties of classical particles or waves emerge exclusively in the classical limit when macroscopic matter is considered.

Not only for photons, but for all matter quanta

Now that light (considered waves before the quantum evolution) share some natures of particles, would matter like electrons, atoms, etc (considered particles before the quantum evolution) share some nature of waves as well?

In 1924, de Broglie proposed that particle-like matter should also have wave-like properties. The de Broglie wavelength of a particle is proposed to be

$$\lambda = h/p \tag{4.3}$$

Though sounds crazy at that time, the de Broglie matter wave is verified in experiments. For example, the interference pattern in the single photon double-slit experiment above is also observed using electrons, atoms, and so on.

> **More success of matter wave**

In fact, crystal diffraction experiments are relatively easier ways to illustrate matter waves. The diffraction pattern of electrons are observed as early as 1927. Also the matter wave postulate immediately inspired Schrödinger to ask the question: what is the wave equation of matter wave? He then derived the Schrödinger equation in 1925 and published it in 1926.

Non–relativistic quantum mechanics

Although quantum mechanics is initiated from photons, photons move at speed of light and thus is relativistic by its nature. The quantum theory consistent with special relativity (quantum field theory) was established much later than the non-relativistic quantum mechanics. This relativistic theory is beyond the scope of the current course.

Thus, now let us switch our attention from photons to massive microscopic particles, such as electrons, atoms, etc. They are non-relativistic at low energies. And our focus is paid on how Newtonian mechanics get generalized into non-relativistic quantum mechanics. We will call non-relativistic quantum mechanics just quantum mechanics for short.

> **(Optional) A field of particles**

The wave nature of particles provides us new insight about whether fundamental particles are the most "fundamental" objects. Think about water wave propagating on water surface. These water waves are not fundamental. They are excitations of the more fundamental water surface. Similarly, particles can be considered as excitations of fields. For example, photons are excitations of E&M field (which you have learned), and electrons are excitations of electron field (which you probably have not learned so far). The quantum theory of studying these fields is known as "quantum field theory", which consistently put special relativity into the framework of a quantum theory.

In this course we restrict our attention to non-relativistic quantum mechanics. At this point it is fine to temporarily consider particles as "fundamental", and study their natures.

The wave and particle properties of quanta

(1) Quantized (particle-like). Quanta are discrete and can be counted.

(2) Superposition (wave-like). Like waves, quanta obey linear equations of motion. One can thus add one solution and another to make a third solution.

(3) Energy and angular frequency (connection of wave and particle). The particle-like energy and the wave-like angular frequency is related by the Planck formula

$$E = \hbar\omega \qquad (4.4)$$

(4) Momentum and wavenumber (connection of wave and particle). The particle-like momentum and the wave-like wavenumber is related by the de Broglie formula (4.3)

$$p = \hbar k \qquad (4.5)$$

We will expand some of these properties in greater detail in the next section, and study a few other wave-like and particle-like properties in later sections.

4.2 The Quantum Wave Function

As we discussed, all matter has wave properties. How to describe wave? We are familiar with plane waves with angular frequency ω and wave number k, it looks like $\exp(\mathrm{i}kx - \mathrm{i}\omega t)$. It is a function of space and time, and a special example of wave function.

The quantum wave function

We assert that a quantum state $|\psi\rangle$ of a particle is described by a wave function. For a quantum particle moving in one spatial dimension, the wave function can be written as $\psi(x,t)$. In general, ψ can take complex values. This wave function completely describes the quantum nature of the particle.

Why bother to introduce a wave function? What's its physical meaning? In the below subsections we will answer these questions. But before to proceed, let me remind you that at a given time, a whole function $\psi(x,t_0)$ contains much more information than the position and momentum of a particle (which are two real numbers). Thus a quantum state contains much more information than a classical state. We will see this again and again later.

> **What is the symbol $|\psi\rangle$?**
>
> Don't be scared by the strange appearance of $|\psi\rangle$. For the moment being, it means nothing more than a state (the status of the quantum particle). In the part of quantum information, you will see why this notation is convenient.

4.2.1 The Wave Function as a Probability Amplitude

The wave function is a probability amplitude

From the single quantum double-slit experiment, we learned that we have to introduce probability into the theory of quantum mechanics in a fundamental way.[1] The wave function is a probability amplitude, such that $|\psi(x,t)|^2$ is the probability density:

The probability to find the particle between x and $x+\mathrm{d}x$ is $|\psi(x,t)|^2\mathrm{d}x$.[2]

This is known as the Born's rule and is the physical meaning of the wave function.

Features of the wave function in probability theory

As a probability amplitude, we can immediately read off a few features of $\psi(x,t)$:
The wave function is normalized as

$$\int_{-\infty}^{\infty} |\psi(x,t)|^2 \mathrm{d}x = 1 \tag{4.6}$$

because it describes a particle, and the probability to find the particle once in the whole space is one.

If we prepare many copies of the same state, and measure the position x of the particle for each copy, the average value (called the *expectation value*) of the position is

$$\langle x \rangle = \int_{-\infty}^{\infty} x|\psi(x,t)|^2 \mathrm{d}x \tag{4.7}$$

---

[1] This explanation is first discovered by Born (1926).

[2] For a complex number, $|\psi|^2 \equiv \psi^*\psi$, where star denotes complex conjugate.

More generally, the expectation value for a function of position $f(x)$ is

$$\langle f(x) \rangle = \int_{-\infty}^{\infty} f(x) |\psi(x,t)|^2 \mathrm{d}x = \int_{-\infty}^{\infty} \psi^*(x,t)\, f(x)\, \psi(x,t)\, \mathrm{d}x \qquad (4.8)$$

Why we want $\langle f(x) \rangle$?

Because $\langle x \rangle$ does not provide us enough information about the distribution. We want to know more. For example, the two distributions are shown in Fig. 4.10:

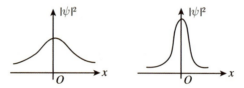

Fig. 4.10

They have the same $\langle x \rangle$. But in the left panel the value of x is less certain because the distribution is broader, while in the right panel we have a better idea where to find the particle. This information is encoded in $\langle x^2 \rangle$.

Remaining questions

We now have already known a lot about a quantum state. However, you may have the following questions:

(1) So far only $|\psi(x,t)|^2$ appears in the description. So why do we use $\psi(x,t)$ as the fundamental object?

(2) How to relate $\psi(x,t)$ to the single quantum double-slit experiment?

(3) We have shown how to extract position information from $\psi(x,t)$. However, we need to extract the momentum information as well to know the state better. Can we and how do we do it?

(4) We have only discussed the x dependence of $\psi(x,t)$, what about the t dependence?

We will address (1) and (2) in section 4.2.2, (3) in section 4.2.3, and (4) in section 4.5.1.

4.2.2 Consequence of Superposition and Linearity

The world is linear

Superposition is the key feature of wave. Mathematically, superposition means linearity: if wave functions ψ_1 and ψ_2 are solutions of the wave equation, then $c_1\psi_1 + c_2\psi_2$ is also a solution, where c_1 and c_2 are arbitrary complex-valued constants.

The water waves and E&M waves are approximately linear when the amplitude is small. However, quantum mechanics asserts that the wave function is *exactly* linear.

> **Why we introduce probability amplitude?**
>
> Now we know why we introduce probability amplitude in quantum mechanics: Because we treat probability as waves. Waves are linear in amplitude, but not in energy or intensity. This is a general feature of waves. For example, the intensity of light and the energy of water waves are both proportional to amplitude squared.

Double-slit and the interference in probability

Suppose $\psi_1(x)$ and $\psi_2(x)$ are the solutions going through silt 1 and slit 2, respectively. Then from linearity, $\psi(x) = \psi_1(x) + \psi_2(x)$ is also a solution. From the symmetry of the double-slit setup, this should be the actual solution that we look for.[①]

At the dark bands, though $\psi_1 \neq 0$, $\psi_2 \neq 0$, they cancel such that $\psi_1 + \psi_2 = 0$.

Thus, if we block a slit (let's slit 2), the dark band is no longer dark since the probability density to have particle at the (originally) dark band is $|\psi_1|^2 \neq 0$. However, if we allow both slits, the probability density becomes $|\psi|^2 = |\psi_1 + \psi_2|^2 = 0$.

> **Global and relative phases**
>
> Here, the relative phase (a phase is $e^{i\alpha}$ for real α) between ψ_1 and ψ_2, in the form $\psi_1 + e^{i\alpha}\psi_2$ can have physical effects. Because the phase controls constructive interference, destructive interference or something in between.

[①] Here we have written $\psi(x,t) \sim \psi(x)$ for a time-independent problem. We will later see that it is indeed consistent. Also, we are not careful about complex phases at this moment.

However, a global phase of the whole wave function ψ, in the form $e^{i\alpha}\psi$ (if we do not do further superpositions or enlarge the system) is unobservable (and thus unphysical). Because a global phase does not affect the probability density $|\psi|^2$.

Linearly adding up probability amplitude ψ is how quantum mechanics works. This is totally different from adding up probabilities in our classical life.

The probabilities are non-negative, and has no interference when adding them up. For example, if we buy a lottery and then buy another. The probability to win simply adds up. The probability of buying one lottery will not cancel the probability of buying the other.

However, adding probability amplitude is different. Because

$$|\psi|^2 = |\psi_1|^2 + |\psi_2|^2 + \psi_1^*\psi_2 + \psi_2^*\psi_1 \qquad (4.9)$$

The first two terms correspond to adding up probabilities. But there are two additional interference terms $\psi_1^*\psi_2 + \psi_2^*\psi_1$. They are not positive definite and can cancel the whole thing.

I hope this relates the single-quantum double-slit experiment to the wave function of a quantum, and explains Alice's observation ⑥ at the beginning of this part.

Plane waves as building blocks of wave functions in free space

In math, linearity allows us to expand the wave function with a complete set of basis. Physically, the importance of basis is that we can first understand these simple states, and then consider a general state as a weighted average of these simple states.

The bases are the solutions of the wave equation and of course depend on detailed problems: for a particle moving in potential $V(x)$, the solutions (and thus bases) depend on the form of $V(x)$. For most of this part (unless explicitly stated), we will be interested in the case of a constant V (or gluing a few stages of motion where each stage has constant V). For constant V, happily the bases of the wave function are the familiar plane waves

$$\psi_p(x,t) = \frac{1}{\sqrt{2\pi\hbar}} e^{i(px-Et)/\hbar} \qquad (4.10)$$

Here we have used $p = \hbar k$ and $E = \hbar\omega$ to replace wave vector and angular frequency with momentum and energy of the quantum. Positive and negative p indicate that the wave is moving to the right and left, respectively. Again, reminding that by wave or oscillation here, there is nothing changing positions periodically as for water wave. But rather, what really oscillates is the *probability amplitude*.

Plane waves are not normalized

We required the ψ being normalized: $\int |\psi(x,t)|^2 dx = 1$. Unfortunately, a plane wave cannot be normalized since this integral diverges for $\psi_k(x,t)$. Physically it is because finding the plane wave at anywhere in space is a constant: $|\psi_k(x,t)|^2 = \text{const}$. Thus the normalization factor (whose square is probability density) is zero. An analogue is that, if you guess an integer from 1 to 10, you can win by chance 1/10. But if you guess a number uniformly distributed from $-\infty$ to ∞, your chance to win is zero.

(1) If you really worry about this normalization, you may choose to do one of the follows:

(2) Start from finite volume of space, and take the volume of space infinity after all calculations.

Consider a wave packet — a collection of plane waves, which locally looks like plane waves but can be normalized properly. Physically, we can never make an exact plane wave.

But they may make things unnecessarily complicated. We thus leave the plane waves unnormalized. The factor $1/\sqrt{2\pi\hbar}$ is not a normalization factor but just for future convenience.

Thanks to linearity, a general state of a quantum can be written as a superposition of the plane waves

$$\psi(x,t) = \int_{-\infty}^{\infty} dp\ c(p)\psi_p(x,t) \qquad (4.11)$$

where $c(p)$ are complex valued functions of p. Here $\psi(x,t)$ can be thought of as a weighted average of plane waves $\psi_p(x,t)$ with weight $c(p)$. And $c(p)$ can be thought of as how much of $\psi_p(x,t)$ is contained in a state $\psi(x,t)$.

Superposition opens up the possibility to consider more exotic objects, such as the Schrödinger's cat (section 4.7) and quantum entanglement (the next chapter).

4.2.3 Extracting Momentum Information from the Wave Function

In Newtonian mechanics, to describe a state, we would like not only to know the position of the particles but also how they move (momentum). Now that a wave function describes the state of a quantum, we also would like to know how to extract the information of momentum from the wave function.

States with certain and uncertain momentum

What is the momentum of a wave function $\psi(x,t)$? We have a few observations as follows:

(1) For plane wave ψ_p, we know that its momentum is p.

(2) For general ψ, we do not expect a definite answer because of superposition. For example, the state can be $\psi = \psi_{p_1} + \psi_{p_2}$ $(p_1 \neq p_2)$. What's the momentum of ψ? The state does not have a definite momentum, but rather we have probabilities to find the state with momentum p_1 and p_2.

Reconsidering normalizations

You may be puzzled here: If I take $\psi = \psi_{p_1}$ and $\hat{p} = p_1$, then isn't $\int \psi_{p_1}^* p_1 \psi_{p_1} \, dx = \infty$? You are right! This is because ψ_{p_1} cannot be normalized.

To fix this, we may reconsider the expectation value and define: $\langle \hat{p} \rangle_r = \langle \hat{p} \rangle / \langle 1 \rangle$, where $\langle 1 \rangle \equiv \int \psi_{p_1}^* \psi_{p_1} \, dx$.

At first sight, this seems stupid: I am trying to make sense of ∞ / ∞. However, in physics, this indeed makes sense: Plane waves in infinite space does not exist in its exact form in nature. This is the origin of the divergence. We thus can first *regularize* the infinity by considering wave packets close enough to plane waves, or plane waves in a finite box. In such physical problems, the infinities are canceled and we get the desired result $\langle \hat{p} \rangle_r = p_1$ for $\psi = \psi_{p_1}$.

Computing expectation value involving momentum

We have shown how to compute an expectation value $\langle f(x) \rangle$ for a function of position of the particle. For the momentum of the particle, we hope to find a similar formula

$$\langle \hat{p} \rangle = \int_{-\infty}^{\infty} \psi^*(x,t) \, \hat{p} \, \psi(x,t) \, dx \qquad (4.12)$$

Is it possible? Here we have put a hat to momentum p. Why? We hope $\hat{p}$ to tell information about momentum. However, we cannot simply take $\hat{p} = p$. Because the RHS does not have p elsewhere in the equation. Thus putting a number p in the integral does not make sense. So how to extract momentum information from the wave function?

For example, if we take a plane wave $\psi = \psi_{p_1}$, we should expect $\hat{p} \psi_{p_1} = p_1 \psi_{p_1}$. If we take $\psi = \psi_{p_2}$, we should expect $\hat{p} \psi_{p_2} = p_2 \psi_{p_2}$. Is this possible for the same $\hat{p}$?

Chapter 4 Quantum Mechanics   115

This suggests that we take $\hat{p}$ as a differential operator (known as the momentum operator), instead of a number:

$$\hat{p} = -i\hbar \frac{\partial}{\partial x} \equiv -i\hbar \partial_x \qquad (4.13)$$

where the ∂_x in the very RHS is just a short hand notation of $\partial/\partial x$. We assert that (4.13) applies not only for free space, but also for non-constant V.

More generally, the expectation value $\langle f(x,\hat{p}) \rangle$ can be extracted from the wave function as

$$\langle f(x,\hat{p}) \rangle = \int_{-\infty}^{\infty} \psi^*(x,t)\ f(x,\hat{p})\ \psi(x,t)\ \mathrm{d}x \qquad (4.14)$$

4.3 Observables and Measurements

When talking about expectation values in the previous section, we implicitly mean measurements: measuring many identical quantum states and take the average. In this section, let us address the question of measurements more carefully. Especially, if we only have one system and do a measurement once, how do we address the probability of measurement outcomes?

We have talked about a key difficulty: When measuring momentum, what happens if the state does not have definite momentum? In this section, we will address the theory of measurements in quantum mechanics: observables, states and measurement outcomes (Fig. 4.11).

Fig. 4.11

> **Classical measurements are simple**

In classical theories, theorists never worried about measurements:

A classical state always have definite values of observables.

A classical measurement can be made in an infinitely gentle way which do not disturb the system.

Quantum systems do not have above features. Thus, measurement is a key part of the quantum mechanics theory.

Observables are represented by operators

To talk about measurements, we should start with what we can measure — observables. For example, position and momentum are observables. In the example of momentum operator $\hat{p} = -i\hbar \partial_x$, we learned that a number may not be enough to represent an observable. For position, x as a number can also be considered as a special operator $\hat{x} = x$. In general:

An observable is represented by an operator $\hat{O}$ in quantum mechanics.

Moreover, for the observable outcome being real numbers instead of complex numbers, we require that the observables are Hermitian operators: $\hat{O}^\dagger \equiv (\hat{O}^*)^\mathrm{T} = \hat{O}$.

States with definite values of observables

Among the quantum states, some states have definite value λ of $\hat{O}$:

$$\hat{O}\psi_\lambda(x) = \lambda \psi_\lambda(x) \qquad (4.15)$$

In other words, the operator acting on the wave function is equal to the number λ acting on the state. For these states, when we measure $\hat{O}$ on the state, we get definite outcome λ. Mathematically, we call the state $\psi_\lambda(x)$ an eigenstate (or sometimes eigenfunction in math) of $\hat{O}$ with eigenvalue λ.

Examples:

(1) Momentum eigenstates: The plane waves are momentum eigenstates with definite

---

① In this section, we will talk about the time around the measurement t_m and suppress the time variable of the wave function for short: $\psi(x, t_m) \to \psi(x)$.

momentum. Indeed, we already know that plane wave ψ_p defined in (4.10), and

$$\hat{p}\psi_p(x) = p\psi_p(x) \tag{4.16}$$

When we measure the momentum of this state, we definitely get p.

(2) Position eigenstates: What kind of states have definite position? Recall the probability amplitude nature of the wave function, the state with definite position q should only take value at one particular position q, and vanish everywhere else. This function is known as a Dirac δ-function defined as[①]

$$\psi_q(x) = \delta(x-q) \equiv \begin{cases} 0, & \text{if } x \neq q \\ \infty \text{ with } \int_{-\infty}^{\infty} \delta(x-q) = 1, & \text{if } x = q \end{cases} \tag{4.17}$$

We can test that $\psi_q(x)$ is indeed an eigenstate of $\hat{x} = x$:

$$\hat{x}\psi_q(x) = q\psi_q(x) \tag{4.18}$$

This is because if $x \neq q$, $\psi_q(x) = 0$. We thus have $x\psi_q(x) = q\psi_q(x)$. When we measure the position of this state, we definitely get q.

For the eigenstates, the outcome of measurement is simple: we get the corresponding eigenvalue once measured. However, what about superpositions, such as two plane waves $\psi_{p_1} + \psi_{p_2}$, or two Dirac δ-functions $\psi_{q_1} + \psi_{q_2}$?

Measurement outcome for a general state

For superpositions when measuring $\hat{O}$, we decompose the general state in eigenstates.

(1) For observables taking discrete results: The decomposition is

$$\psi(x) = \sum_i c_i \psi_{\lambda_i}(x) \tag{4.19}$$

where ψ_{λ_i} is the eigenstate of $\hat{O}$ with eigenvalue λ_i. The probability for the measurement outcome being λ_i is $|c_i|^2$. After the measurement, the wave function "collapse" to ψ_{λ_i}.

(2) For observables taking continuous results: The decomposition is

$$\psi(x) = \int c(\lambda)\psi_\lambda \, d\lambda \tag{4.20}$$

where ψ_λ is the eigenstate of $\hat{O}$ with eigenvalue λ. The probability density for the measurement outcome being λ is $|c(\lambda)|^2$. After the measurement, the wave function "collapse" to ψ_λ.

[①] Unfortunately, ψ_q cannot be properly normalized either. Again, if you'd like to properly normalize ψ_q, you can either study particle in a finite box, or a physical narrow peak instead of the δ-function.

Examples:

(1) Momentum measurement: Eq. (4.11) realizes Eq. (4.20). After the measurement, the probability density to find p is $|c(p)|^2$.

(2) Position measurement: A state $\psi(x)$ can always be written as

$$\psi(x) = \int_{-\infty}^{\infty} \psi(q)\delta(x-q) \, \mathrm{d}q \tag{4.21}$$

After the measurement, the probability density to find q is $|\psi(q)|^2$. This reproduces the Born's rule, i.e. the probability amplitude interpretation of a quantum wave function.

"Collapse" of wave function?

What does "collapse" of a wave function after a measurement mean? We mean the wave function becomes the corresponding eigenstate as indicated by the meaning of the word "collapse". However, currently there is no unique understanding about dynamically why and how the collapse happen. This question is related to the interpretation of quantum mechanics. We will address it briefly in the epilogue section.

At the beginning of this part, when Alice is walking in the brightness, her position keeps being measured. Thus she knows better where she is. This explains ②.

Wave Function Collapse Algorithm

Interestingly, the collapse of the quantum wave function has inspired the classical programming world. The Wave Function Collapse Algorithm is developed to dynamically generate infinitely sized 2D or 3D maps. The algorithm is inspired by the quantum superposition and collapse of the wave function.

4.4 The Uncertainty Principle

We have lots of uncertainties in our classical life due to lack of information. For example, before your exam score is released, your teacher tells you only the mean and standard deviation (uncertainty) of score distribution. What can you tell from this information? If the standard deviation is small, you are more certain about your score, but less certain about your ranking (as a small mistake may bring you from top to below mean). If the standard deviation is large, you have less idea about your score but can estimate better your ranking based on usual performance.

In the quantum world, even if we have complete knowledge of the state, we still have uncertainties. For example, what happens if you measure the position of a plane wave?

In this section, let us make this question better defined and more general.

Uncertainty is quantified by standard deviation

Talking about precision (or uncertainty), the corresponding quantity in statistics is the standard deviation. Given a state:

(1) Uncertainty of the state's position σ_x is $\sigma_x \equiv \sqrt{\langle (x - \langle x \rangle)^2 \rangle} = \sqrt{\langle x^2 \rangle - \langle x \rangle^2}$.

(2) Uncertainty of the state's momentum σ_p is $\sigma_p \equiv \sqrt{\langle (\hat{p} - \langle \hat{p} \rangle)^2 \rangle} = \sqrt{\langle \hat{p}^2 \rangle - \langle \hat{p} \rangle^2}$.

The uncertainty principle

In math, a Kennard inequality tells that for $\hat{p} = -i\hbar \partial_x$:

$$\sigma_x \sigma_p \geq \frac{\hbar}{2} \tag{4.22}$$

This is the uncertainty principle, first discovered by Heisenberg in 1927.

We are trying to pack all the difficulties into math.[①] Instead of proving (4.22), let us see some examples.

Understanding the uncertainty principle from waves

Let's see a few examples of wave functions:

(1) Plane waves: $\psi_p = \dfrac{1}{\sqrt{2\pi\hbar}} e^{ipx/\hbar}$, $\sigma_p = 0$ and $\sigma_x = \infty$.

[①] And going through the math proof does not make me feel it being more intuitive. That's why I choose to hide the math here and leave it for a proper quantum mechanics course. You will find it more intuitive once you realize that $\psi(x)$ and $c(p)$ are Fourier transform to each other. For Fourier transformation in general, the original function and the image admits an uncertainty principle. But this is beyond the scope of this course.

(2) Gaussian wave packet:

$$\psi(x) = \frac{1}{\sqrt{\sigma_x \sqrt{2\pi}}} e^{-\frac{x^2}{4\sigma_x^2}} \quad (4.23)$$

It's simple to check that σ_x is indeed the standard deviation for x. To calculate σ_p, noting that $\langle \hat{p} \rangle = 0$ and $\langle \hat{p}^2 \rangle = \hbar^2/(4\sigma_x^2)$, thus $\sigma_p = \hbar/(2\sigma_x)$. The Gaussian wave packet exactly saturates the uncertainty principle.

As the wave packet does not have definite momentum, different momentum eigenstates in the superposition do not move equally fast. That indicates that the wave packet will spread over time. We will see that it's indeed the case in section 4.5.1.

Understanding the uncertainty principle from particles

Can we first measure x and then measure $\hat{p}$ for the same state, and thus get both position and momentum information precisely?

The quick answer is no, because after a quantum measurement, the state collapses to the eigenstate. But this argument still seems mysterious. Let us consider a scenario in which we can see what actually happens.

Any measurement must act some interaction on the system and thus must change the system.

Fig. 4.12

For example, you see an apple because light is reflected by the apple and is detected by your eyes (Fig. 4.12). The light pushes the apple at the same time.

Classically, the reaction on the system can be made as small as one wants. However, quantum mechanically it is not possible.

To measure the position precisely, one needs to use a wave packet of light with shorter wavelength. We at best have $\sigma_x > \lambda$.

From de Broglie's relation, $p_\lambda \sim \hbar/\lambda$ and thus shorter wavelength implies larger p_λ. The reflection of light transfers momentum of order p_λ to the measured object. Thus the object has a momentum uncertainty of order $\sigma_p \sim p_\lambda$. As a result, we at least need $\sigma_x \sigma_p \sim \hbar$.

Wave vs particle viewpoints

There is a subtle difference between understanding the uncertainty principle from waves and particles: From waves, we are talking about the intrinsic fuzzyness of quantum states.

This is more precise in the sense of a princple of uncertainty. From particles, we see that fundamentally when measuring a microstate's position we have to change its momentum. These two aspects are nevertheless consistent with each other.

(Optional) The quantum non-cloning theorem

You may think of another way to invalidate the uncertainty principle: What about cloning the state into two copies, and measuring position and momentum, respectively?

Unfortunately, your brilliant idea has been elaborately blocked by the theory of quantum mechanics, by the quantum non-cloning theorem — no machine can clone an unknown state without destroying the original copy.

To prove that, we need to extend our language a little: including states of a cloning machine and multiple outcome states. Also, it is convenient to use the abstract notation $|\text{object}\rangle$ to denote the quantum state of the object. If a quantum cloning machine $|M\rangle$ exists, it does the follows: for any given quantum state $|\psi\rangle$, the machine transforms the state as

$$|M\rangle|\psi\rangle \to |M_\psi\rangle|\psi\rangle|\psi\rangle \qquad (4.24)$$

where $|M_\psi\rangle$ is the state of the machine after the transformation. Now, consider two states $|1\rangle$ and $|2\rangle$. We expect

$$|M\rangle|1\rangle \to |M_1\rangle|1\rangle|1\rangle, \qquad |M\rangle|2\rangle \to |M_2\rangle|2\rangle|2\rangle \qquad (4.25)$$

Now what about the state $\alpha|1\rangle + \beta|2\rangle$?

(1) From how a cloning machine should work, we should get $|M_*\rangle(\alpha|1\rangle + \beta|2\rangle)^2$.

(2) From linearity of quantum mechanics, we should get $\alpha|M_1\rangle|1\rangle|1\rangle + \beta|M_2\rangle|2\rangle|2\rangle$.

Although we don't know the final state of the machine $|M_1\rangle$, $|M_2\rangle$ and $|M_*\rangle$, the above (1) and (2) cannot be consistent. Because (1) contains $\alpha\beta|1\rangle|2\rangle$ which is not part of (2). We thus conclude that a quantum cloning machine is impossible. Linearity is the key to prove the non-cloning theorem.

(Optional) what is possible about cloning

It's important to note the conditions of the non-cloning theorem.

It is possible to produce many copies of a known state. However, the ability of producing a known state does not help you to measure the position and momentum of an unknown state.

It is possible to clone a state and at the same time destroy the original copy. Observiously this does not help you to invalidate the uncertainty principle either.

> **The secure quantum information**
>
> Thanks to the non-cloning theorem, if Alice send to Bob a message by information carried on single quanta, once a spy Charlie wiretaps the information, the quanta is gone and Bob will no longer get the information. Bob can then alert Alice stop sending the information. Thus, quantum information has built-in security in communication proven by math.
>
> But Charlie has still got some information. Secret still leaked a bit, right? One can avoid that by not sending the actual information, but sending the key to decrypt information instead. The key is useless without the actual message. If Bob cannot get the key, Alice will send a new key. Even Charlie wiretaped and got an expired key, it is useless.

A joke says when Heisenberg knows where he is, he doesn't know how fast he is walking. So does Alice for item ① at the beginning of this part.

4.5 The Schrödinger Equation

We have talked about the interpretation of quantum states and measurements. They belong to the properties of the state at a given time.

But more importantly, physics is about given an initial condition, how to predict the state after some time. In other words, it is important to know the equation of motion which governs the time evolution of the system. In quantum mechanics, this governing equation is the Schrödinger equation.

4.5.1 The Schrödinger Equation

Extracting energy from a wave function: the Schrödinger equation

We discussed in section 4.2.3 that the operator $-i\hbar\partial_x$ can extract momentum from a plane wave $\psi_p(x,t) \propto \exp[i(px - Et)/\hbar]$, and consequently any superposition of plane waves, and thus a general wave function.

How to extract the information of energy from a wave function? We can do it in two ways:

(1) Using the relation $E = \dfrac{p^2}{2m} + V(x)$ and replacing $p \to \hat{p}$.

(2) Noting that applying $i\hbar\partial_t$ on plane waves directly extract their energies. These two ways must be consistent. We thus get an equation

$$i\hbar\partial_t \psi(x,t) = \hat{H}\psi(x,t) \ , \qquad \hat{H} \equiv -\frac{\hbar^2 \partial_x^2}{2m} + V(x) \qquad (4.26)$$

This is the Schrödinger equation, and $\hat{H}$ is called the Hamiltonian (the operator corresponding to energy). The LHS connects the state of a given time to a later time by the appearance of ∂_t.

In this part, we have been talking about one-dimensional problems. For future reference, the Schrödinger equation with three spatial dimensions takes the form

$$i\hbar\partial_t \psi(\boldsymbol{x},t) = \hat{H}\psi(\boldsymbol{x},t) \ , \qquad \hat{H} \equiv -\frac{\hbar^2 \nabla^2}{2m} + V(\boldsymbol{x}) \qquad (4.27)$$

We are not deriving it

Here by arguments and extrapolations, we show that the Schrödinger equation is a natural thing to expect. But we are not deriving the Schrödinger equation. Rather, the Schrödinger equation is a fundamental postulate of quantum mechanics.

The Schrödinger equation appears to be derived here because we have cheated: The plane waves are the quantum state in free space with $V = $ const (even for constant V, we only argued that the plane waves are the natural thing to expect). But now, we are extrapolating to include non-constant V.

The spreading wave packet

We argue that wave packets spread because its uncertain momentum. How to see this in calculations?

Let us focus on the simplest case: $V = 0$. One can verify that the following wave function satisfies the Schrödinger equation:

$$\psi(x,t) = \sqrt{\frac{\sigma}{\sqrt{2\pi}(\sigma^2 + i\beta t)}} e^{i(kx - \omega t)} e^{-\frac{(x - v_g t)^2}{4(\sigma^2 + i\beta t)}} \qquad (4.28)$$

Where $v_g \equiv \hbar k/m$, $\beta \equiv \hbar/(2m)$ and $\omega \equiv \hbar k^2/(2m)$. Here σ is a free parameter, indicating the spatial spread of the wave function at $t = 0$.

This solution is known as the Gaussian wave packet because the probability density is

$$|\psi(x,t)|^2 = \frac{\sigma}{\sqrt{(2\pi)(\sigma^4+\beta^2 t^2)}} e^{-\frac{\sigma^2(x-v_g t)^2}{2(\sigma^4+\beta^2 t^2)}} \qquad (4.29)$$

This is indeed a Gaussian wave packet at any time. The spatial spread of the wave function is

$$\sigma_x(t) = \sqrt{\sigma^2 + \frac{\beta^2 t^2}{\sigma^2}} \qquad (4.30)$$

Why spreading?

Why the wave packet is spreading? Intuitively, a wave packet contains waves with different momenta. These waves travel at different speeds. Thus, after a while, the superposition disperses and the wave packet spreads.

The stationary state Schrödinger equation

Solutions of the Schrödinger equation with definite energy E is of particular interest. This is because:

(1) Physically, energy is conserved. If we start with a state with definite energy and keep it isolated, it will continue to have definite energy.

(2) Mathematically, with definite energy, the Schrödinger equation is greatly simplified. In general, the Schrödinger equation (4.26) is a partial differential equation. For a state with definite energy, we can solve the time part trivially and reduce it to an ordinary differential equation in the x direction only.

For a state with definite E, the Schrödinger equation breaks into two parts:

(1) The time part is $i\hbar \partial_t \psi(x,t) = E\psi(x,t)$. This can be solved as $\psi(x,t) = e^{-iEt/\hbar}\psi(x)$. Thus the time dependence in the wave function is a simple phase.

(2) The Hamiltonian part is

$$\left[-\frac{\hbar^2 \partial_x^2}{2m} + V(x)\right]\psi(x) = E\psi(x) \qquad (4.31)$$

This equation is known as the stationary state Schrödinger equation.

The stationary state Schrödinger equation (4.31) is a powerful tool to explore the quantum wonderland. In the reminder of the section, we will make use of the stationary state Schrödinger equation to consider various types of potentials and how the wave function $\psi(x)$ behaves in these cases.

4.5.2 A Step in the Potential

Here and in subsequent subsections, we focus on square potentials where the potential is a constant at most places, except having a sudden change at the connection points. For such potentials, plane waves still work at almost everywhere, except that at connection points, continuous conditions have to be imposed. This avoids the math difficulty of solving the Schrödinger equation as a differential equation.

The step potential

Consider the potential in the stationary state Schrödinger equation (4.31):

$$V = \begin{cases} V_1, & \text{if } x < x_0 \\ V_2, & \text{if } x > x_0 \end{cases} \quad (4.32)$$

Labels A, B, C, D in Fig. 4.13 will be used in the subsection below.

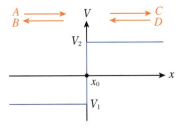

Fig. 4.13

Realization of a step

For example, a charged quanta travelling from zero electric potential to non-zero electric potential realizes a step potential if the transition is sharp (Fig. 4.14).

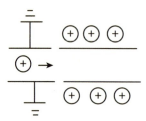

Fig. 4.14

Wave function away from the step

For the step potential, when $x \neq x_0$, V is a constant. The stationary state Schrödinger equation reduces to

$$-\frac{\hbar^2}{2m}\partial_x^2 \psi(x) = (E-V)\psi \qquad (4.33)$$

It has very simple solutions. It is the same plane wave solutions which have motivated us to "derive" the Schrödinger equation:

$$\psi(x) = \begin{cases} Ae^{ik_1 x} + Be^{-ik_1 x}, & \text{if } x < x_0 \\ Ce^{ik_2 x} + De^{-ik_2 x}, & \text{if } x > x_0 \end{cases} \qquad (4.34)$$

where A, C denote waves moving to the right, B, D denote waves moving to the left, and

$$k_1 \equiv \frac{1}{\hbar}\sqrt{2m(E-V_1)}, \quad k_2 \equiv \frac{1}{\hbar}\sqrt{2m(E-V_2)} \qquad (4.35)$$

The connection conditions

What happens to the point x_0? There must be two relations between A, B, C and D. This is because, mathematically, Eq. (4.34) is a second order differential equation and thus should have only two integration constants. Physically, if we set $D=0$, then it becomes a problem of an incoming wave A scatter at the potential and thus B and C should be fixed. Thus there are two relations between A, B, C and D.

The relations are

(1) $\psi(x)$ is continuous. Otherwise, $\partial_x \psi(x_0) \to \infty \Rightarrow$ infinite momentum $\Rightarrow$ unphysical.

(2) $\psi'(x)$ is continuous. Otherwise, $\partial_x^2 \psi(x_0) \to \infty \Rightarrow$ infinite energy $\Rightarrow$ unphysical.

(2) $\psi'(x)$ is continuous. Otherwise, $\partial_x^2 \psi(x_0) \to \infty \Rightarrow$ infinite energy $\Rightarrow$ unphysical.
Applying these two relations to connect the two branch of solutions in Eq. (4.34):

$$A e^{ik_1 x_0} = \frac{k_1 + k_2}{2k_1} C e^{ik_2 x_0} + \frac{k_1 - k_2}{2k_1} D e^{-ik_2 x_0},$$
$$B e^{-ik_1 x_0} = \frac{k_1 - k_2}{2k_1} C e^{ik_2 x_0} + \frac{k_1 + k_2}{2k_1} D e^{-ik_2 x_0} \quad (4.36)$$

Scattering into classically allowed region with $E > V_1$ and $E > V_2$

As we mentioned, when setting $D = 0$, the problem is reduced to a scattering problem. A is the incoming wave coming from $x \to -\infty$, B is the reflection wave and C is the outgoing transmission wave.

Here we first consider the case when the energy of the state is large enough, such that if the state was a classical particle, then it is able to reach the $x > x_0$ regime.

From Eq. (4.36) we get

$$C = \frac{2k_1}{k_1 + k_2} e^{i(k_1 - k_2)x_0} A$$
$$B = \frac{k_1 - k_2}{2k_1} e^{i(k_1 + k_2)x_0} C = \frac{k_1 - k_2}{k_1 + k_2} e^{2ik_1 x_0} A \quad (4.37)$$

In general, there are transmission and scattering waves. This is different from the particle mechanics that the particle always goes through (because it has enough energy).

Scattering into classically forbidden region with $V_1 < E < V_2$

Now we consider the case $E < V_2$. In this case, classically the particle is unable to reach the $x > x_0$ regime.

In this case, k_2 becomes imaginary. When $x > x_0$,

$$\psi(x) = C e^{-|k_2|x} \quad (4.38)$$

We are forced to have $D = 0$ because the D term blows up at $x \to \infty$. The formal solution of C is the same as Eq. (4.37).

How the particle can enter the classical forbidden regime even if the $V_2 > E$ (Fig. 4.15)? This is because formally the effective "kinetic energy" $\hbar^2 k_2^2/(2m) < 0$ when k_2 is imaginary.

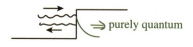

Fig. 4.15

An exponentially small part of wave function can enter the classical forbidden regime. This is completely different from classical particles and has profound implications[①]. We shall uncover some of them later, namely tunneling, a consistent picture of identical particles.

4.5.3 The Potential Barrier: Reflection and Tunneling

Now we modify the step potential further: Let's join two steps to make a potential barrier.

Scattering on a potential barrier

Consider the potential barrier illustrated in Fig. 4.16. The detailed calculation will be left as an exercise. And here we shall use the experience we got from the step potential to study qualitative features here.

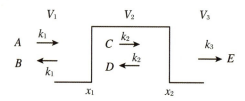

Fig. 4.16

The way of solving this problem is similar to the last subsection: In principle we just have to solve an array of 4 equations at x_1 and x_2, to determine 4 relations between 5 coefficients.

Practically, it is convenient to start from the outgoing wave E. Given E, we get C and D. And given C and D, we get A and B. In other words, given non-vanishing A, we always have non-vanishing E, even in the case that $E < V_2$.

[①] However, in the wave perspective, it is not that hard to understand — E&M wave into a conductor has similar exponentially decaying properties (instead of vanish immediately).

The quantum tunneling effect

In the case $E < V_2$, classically the particle can never go from A to E because there is a barrier to block it (Fig. 4.17).

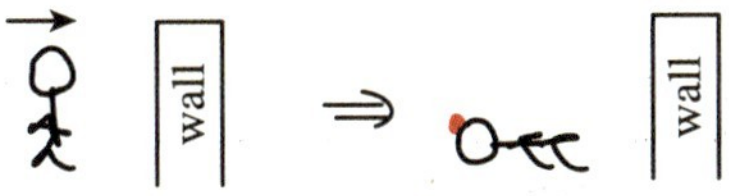

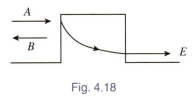

Fig. 4.17

However, in quantum mechanics, no matter how high the barrier is (in the real world the barrier is always finite), there is an exponentially small part which goes into the barrier and escapes out to E (Fig. 4.18)①. This is a key feature of quantum mechanics.

Fig. 4.18

Example of tunneling: α-decay

Heavy elements may be unstable. They may emit an α particle, i.e. the Helium nuclei $^4_2$He and the remaining part becomes another element. This process is known as the α-decay.

For example, the following process can happen:

$$^{238}_{92}\text{U} \to {}^{234}_{90}\text{Th} + {}^4_2\text{He} \tag{4.39}$$

Here, the $^{234}_{90}$Th part of the nuclei provides a binding potential for the $^4_2$He part of the nuclei. They are together to form the $^{238}_{92}$U. But the $^4_2$He has a small chance to escape, and this is the α-decay of $^{238}_{92}$U (Fig. 4.19).

The element $^{238}_{92}$U has a half-life of 10^{17} s $\sim 4 \times 10^9$ a (about 1/3 of the age of the universe). Note that the time scale of a typical microscopic process is often a tiny fraction of a second. Such a huge difference in time scales is because of the exponentially small decay rate.

---

① This is why at the beginning of this part, the quantum Alice observe that she can sometimes walk into a wall to enter a room in item ④.

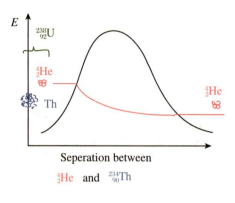

Fig. 4.19

4.5.4 The Potential Well: Scattering and Bound States

What about to connect the two steps differently, to form a potential well?

The potential well

Consider a potential well as in Fig. 4.20. The energy of the quantum classifies the problems into two cases:

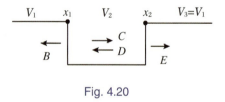

Fig. 4.20

(1) High energy scattering problem: If $E > V_1$ and $E > V_3$, the situation is similar to the discussion in the last subsection.

(2) Low energy bound state: What if $E < V_1$ and $E < V_3$? There is no wave coming from $x \to -\infty$ or wave going towards $x \to \infty$. As a result, the quantum state is confined inside (or a little bit outside with exponentially decaying amplitude) the potential well.

Solving the bound state problem

Recall that for the wave function and its derivative to be continuous at x_1 and x_2, there are 4 conditions. But now there are only 4 variables B, C, D and E and thus there should be at most 3 relations between the 4 variables. What would be the missing condition?

To see that, we apply the continuity conditions for $\psi(x_1)$, $\psi'(x_1)$ and $\psi(x_2)$, $\psi'(x_2)$ explicitly. At x_1,

$$\frac{k_1+k_2}{2k_1}Ce^{ik_2x_1} + \frac{k_1-k_2}{2k_1}De^{-ik_2x_1} = 0 \tag{4.40}$$

$$\frac{C}{D} = \frac{k_2-k_1}{k_2+k_1}e^{-2ik_2x_1} \tag{4.41}$$

Similarly (we can do the replacement $k_2 \to -k_2$, $k_1 \to -k_1$, $x_1 \to x_2$, $C \leftrightarrow D$), at x_2,

$$\frac{C}{D} = \frac{k_2+k_1}{k_2-k_1}e^{-2ik_2x_2} \tag{4.42}$$

They must be consistent and thus

$$e^{2ik_2(x_2-x_1)} = \left(\frac{k_2+k_1}{k_2-k_1}\right)^2 = \left(\frac{1+i|k_1/k_2|}{1-i|k_1/k_2|}\right)^2 \tag{4.43}$$

Recall that $k_1 = \sqrt{2m(E-V_1)}/\hbar$, $k_2 = \sqrt{2m(E-V_2)}/\hbar$. Thus the above relation is a requirement on the energy of the state: Only state with such energies can exist. Thus the energy takes a series of discrete values until $E > V_1$, after which there is a continuous spectrum.

Infinite height potential well

Take the limit $V_1 \to \infty$, keeping E and V_2 fixed. In this case, $k_1 \to i\infty$ and as a result, the wave function no longer enters $x < x_1$ or $x > x_2$. In terms of Eq. (4.43), we now have

$$e^{2ik_2(x_2-x_1)} = 1 \quad \to \quad k_2(x_2-x_1) = n\pi \tag{4.44}$$

In this case, the left going and right going waves combines into a sine function

$$\psi(x) = 2iCe^{ik_2x_1}\sin[k_2(x-x_1)] \propto \sin\left[n\pi\left(\frac{x-x_1}{x_2-x_1}\right)\right] \tag{4.45}$$

It is nothing but the standing wave solutions in classical mechanics.

What values can n take?

Taking negative n corresponds to flipping the sign of k_2 and just add a negative sign to ψ.

However, can $n = 0$?

No. Because we start with a particle in the system. If $n = 0$, $\psi(x) = 0$ and nowhere can we find the particle. Thus we can only have $|n| = 1, 2, 3, \ldots$. The energies of these states are

$$E_n = \frac{(n\pi\hbar)^2}{2m(x_2-x_1)^2} + V_2 \tag{4.46}$$

The ground state

Let us focus on the $n = 1$ state. Recall $n \neq 0$. Thus the $n = 1$ state has the lowest energy among all states. This state acts as a "ground" in the energy spectrum and thus is known as the ground state.

The ground state is of particular interest in quantum systems.① This is because typically a quantum interact with other objects. For example, an electron can emit photons and lower its energy. But once the electron reaches its ground state, the electron can no longer emit photons and thus is stable (assuming the potential does not change). Naively, we may have expected that the ground state has the same energy as V_2, since V_2 looks like the "ground" in the potential. However, the ground state energy is in fact higher than V_2.

Why? Since we know that the particle is inside the well ($\sigma_x < x_2 - x_1$), uncertainty principle tells that the momentum of the particle cannot be zero. The momentum adds some kinetic energy to the system.

If we make the potential well narrower, i.e. make $x_2 - x_1$ smaller, then the ground state energy E_1 increases. This is because more uncertainty in momentum is needed.

Bound states in general

Here we only calculated the case of square potential. And only considered infinite potential well in details. However, the qualitative conclusion is general: If the quantum state does not have enough energy to reach infinity, then the state is a bound state localized in or close to the well (Fig. 4.21). The energy of the state can only take discrete values. There exists a lowest energy state known as the ground state. For example, a nuclei has positive charge and creates a set of bound states that electrons can occupy.

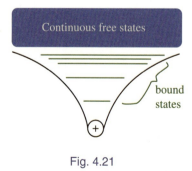

Fig. 4.21

---

① The zero-point energy is an intrinsic feature of quantum mechanics. Neither classical particles nor classical waves has this feature. This is unlike many other features that we have learned. For example, tunneling is already familiar in classical wave mechanics, and bound state is already familiar in classical particle mechanics.

(Optional) Two potential barriers together and resonant tunneling

What about putting two identical potential barriers together? We then have a potential well in between two potential barriers. The potential well can hold bound states in it.

Consider the tunneling problem with such double barrier. Usually, we get double exponential suppression factors. Thus there is doubly small chance to tunnel through double barrier as expected. The situation is illustrated by the gray dashed line in the Fig. 4.22.

Fig. 4.22

However, if we fine-tune the incoming energy to coincide with a bound state energy, then the gray dashed line is not a correct boundary condition for a bound state (for a bound state, both sides should exponentially decay) thus cannot be right. Rather, there is no exponential suppression factor for such tunneling. This situation is known as resonant tunneling.

This explains Alice's enhanced tunneling rate (item ⑤ at the beginning of this part) if we consider the front and back wall as two potential barriers and if Alice's incoming energy coincides with a bound state in the room.

(Optional) Two potential wells together and split of ground state

What about putting two identical potential wells together (called "double well", Fig. 4.23)? Each potential well has its own set of bound states. Do the two set of bound states affect each other?

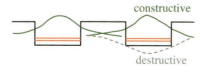

Fig. 4.23

For definiteness, let us call the ground state of the left well ψ_L, and the ground state of the right well ψ_R. Both of them have energy E. Are ψ_L and ψ_R the lowest energy state of the system?

Due to linearity of quantum mechanics, we know that the superpositions $\psi_\pm = (\psi_L \pm \psi_R)/\sqrt{2}$ are also solutions of the Schrödinger equation. In fact ψ_+ has slightly lower energy than E because the wave function takes slightly larger value in the middle barrier (constructive interference), which makes the position of the particle less certain. Similarly, ψ_- has slightly higher energy due to destructive interference. Thus the original ground state splits into two states, with ψ_+ the true ground state of the theory.

(Optional) Infinite potential wells together and the band theory

What about putting infinitely many identical potential barriers (or wells) together? We get a periodic potential. From the experience of double well, the sign of each local ground state (of an individual well) can make constructive or destructive interferences. There are infinitely many choices. Thus the original ground state splits into an infinite number of states. These states form a continuous band (Fig. 4.24).

Fig. 4.24

The particle energy has to be within the band. And resonant tunneling happens and the quanta can freely travel through potential barriers without suppression.

Here we are talking about the ground state. But for other higher energy states, the same argument applies that the bound state energies are broadened into continuous bands (illustrated by the orange thick lines in the figure).

The infinite potential wells can be considered as a toy model of a solid. The potential wells are those made by atoms, and the quanta being studied are the electrons. In solid, the energy of the active electrons is determined by statistical physics and the nature of the solid (the "Fermi energy" at low temperature). If the electrons can move in the band, the solid can conduct electric current and thus is a conductor. If the conducting band is fully occupied, the solid is an insulator. The case in between is known as semi-conductor.

Quantum mechanics in your phone

You should appreciate quantum mechanics for allowing you to have a cell phone. Indeed, the CPU of your phone is doing classical computing, not quantum computing yet. But the logical gates of a modern classical computer is based on semi-conductors, and thus the band theory of solid state physics. Without quantum mechanics, we cannot understand these theories and there is nothing to guide us for making transistors from semi-conductors.

4.6 Identical Particles

At the beginning of this part, item ⑦ of Alice's adventure is that she fails to distinguish an electron from another. In the classical world, we can distinguish persons. Why this fails to work in Alice's quantum adventure?

The classical ways of distinguishing particles all fail

Classically, how we distinguish two persons?

(1) Intrinsic identification: Classically, two persons look different. However, for microscropic particles, this fails to work. Because:

Elementary particles: We have only discovered finite types of elementary particles (labeled by mass, spin and charge). Two particles of the same type have the same intrinsic features. For examples, you cannot find any difference between two electrons.

Composite particles (such as atoms): They are bound states. Bound state energies are discrete. For example, if two hydrogen atoms are both in their ground states, or both in first excited states, we cannot find any difference between the two atoms.

(2) Extrinsic identification: Classically, two persons are separate. We can follow their trajectories to distinguish them. If Bob is in Beijing, and Charlie is in New York, it's easy to confirm their identities even if they look alike.

However, quantum particles can never be in totally different positions. Fundamentally, what separate quantum particles are potential wells. Quantum tunneling tells that the wave function of two particles, say electrons, will always overlap, though the amount may be extremely small. Two electrons are never completely separated.

Those failure of distinguishing quantum particles lead to a fundamental principle of quantum mechanics: the existence of identical particles.

Classification of quantum particles

To make use of the identical particle postulate, we need a wave function to describe at least two particles. The two-particle wave function $\Psi(x_1, x_2, t)$ means that the probability to find one particle at $(x_1, x_1 + dx_1)$ and the other particle at $(x_2, x_2 + dx_2)$ is $|\Psi(x_1, x_2, t)|^2 dx_1 dx_2$.

Since the two particles are identical, we should have $|\Psi(x_1, x_2, t)|^2 = |\Psi(x_2, x_1, t)|^2$. Otherwise, we can tell the difference of the two particles by noticing the probability difference that the two particles can be found.① Thus, $\Psi(x_1, x_2, t)$ and $\Psi(x_2, x_1, t)$ differ at most by

$$\Psi(x_2, x_1, t) = e^{i\alpha} \Psi(x_1, x_2, t) \tag{4.47}$$

where α is a real number. Mathematically, this implies (since x_1 and x_2 are just labels that can be swapped)

$$\Psi(x_1, x_2, t) = e^{i\alpha} \Psi(x_2, x_1, t) \tag{4.48}$$

Thus, $e^{i\alpha} = \pm 1$. The multi-particle wave functions are either symmetric or asymmetric under interchange of particles.

There are thus two kinds of multi-particle wave functions which satisfies the identical particle postulate:

(1) Symmetric wave functions with $\Psi(x_1, x_2, t) = \Psi(x_2, x_1, t)$. The particles described by such wave functions are called bosons. For elementary particles, bosons represents the forces of nature, for example, photons and gravitons.

(2) Anti-symmetric wave functions with $\Psi(x_1, x_2, t) = -\Psi(x_2, x_1, t)$. The particles described by such wave functions are called fermions. For elementary particles, fermions represent the buliding blocks of matter, for example, electrons and quarks.

Pauli's exclusion principle

For fermions, because of the anti-symmetry of the wave function, the fermions cannot be in the same state. This is because, if two fermions were in the same state, then

$$\Psi(x_1, x_2, t) = \Psi(x_2, x_1, t)$$

Considering the definition of fermions

$$\Psi(x_1, x_2, t) = -\Psi(x_2, x_1, t)$$

---

① Imagine one twin likes more to go to a coffee shop and the other twin likes more to go to a bookstore. You can then distinguish them in a statistical sense, and they are no longer identical.

we get $\Psi(x_1, x_2, t) = 0$ and thus such systems do not exist.

This is why in a multi-electron atom, electrons do not only occupy the lowest energy states, but also other states if some electrons have already occupied the lowest energy states.

4.7 Epilogue: Summary and What's Next

In the journey of constructing the theory of quantum mechanics, we have met many difficulties. The difficulties are resolved sometimes by applying what we know; and sometimes by introducing new postulates. Now, let us summarize the postulates that we really have to introduce.

Summary of fundamental postulates of quantum mechanics

(1) The superposition principle of $\psi(\boldsymbol{x}, t)$: the world is linear.

(2) Momentum can be represented by an operator $\hat{\boldsymbol{p}} = -i\hbar\boldsymbol{\partial_x}$.

(3) A measurement is represented by a Hermitian operator $\hat{O}$. The operator $\hat{O}$ defines a complete set of eigenstates $\hat{O}\psi_\lambda(\boldsymbol{x}) = \lambda\psi_\lambda(\boldsymbol{x})$. The state to be measured $\psi(\boldsymbol{x})$ can be expanded by these eigenstates as $\psi(\boldsymbol{x}) = \int c(\lambda)\psi_\lambda \, d\lambda$. The probability density to get value λ from the measurement is $|c(\lambda)|^2$ and the state collapses to $\psi_\lambda(\boldsymbol{x})$ after the measurement.[①]

(4) The wave function obeys the Schrödinger equation

$$i\hbar\partial_t\psi(\boldsymbol{x}, t) = \hat{H}\psi(\boldsymbol{x}, t) , \qquad \hat{H} \equiv -\frac{\hbar^2\nabla^2}{2m} + V(\boldsymbol{x}) \qquad (4.49)$$

(5) Particles can be identical in quantum mechanics.

These postulates do not look as simple as those of relativity. This is why it took decades for the sages of the past century to understand the principles of quantum mechanics. Fortunately, once we have understood and accepted these postulates, the quantum world works elegantly and amazingly.

[①] Recall that if the observable takes discrete values, the corresponding decomposition is

$$\psi(\boldsymbol{x}) = \sum_i c_i \psi_{\lambda_i}(\boldsymbol{x})$$

and the probability to get λ_i is $|c_i|^2$

We have discussed how these postulates are natural guesses inspired from experiments, and explored the consequences of these postulates. Having that said, we do not know what "really happens" in quantum mechanics, especially when a measurement is made. Let me mention it further here:

The Schrödinger's cat: how to understand superposition and measurements?

Traditionally, the Copenhagen (a city where a large part of quantum mechanics is developed) interpretation encourage "shut up and calculate". However, we may not like to shut up at a thought experiment known as the Schrödinger's cat (Fig. 4.25):

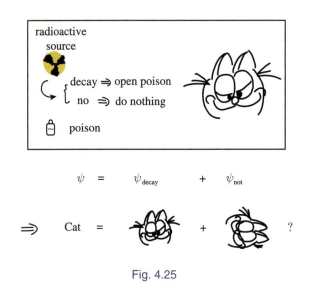

Fig. 4.25

If a quantum process controls whether a bottle of poison opens. And we put a quantum generator, a bottle of poison and a cat in a box. Before we open the box and measure the state, what happens to the cat? Is the cat in a superposition of alive and death?

Unfortunately, no firm answer can be given at the moment. The answers differ in different interpretations of quantum mechanics. In some interpretations, one may even so crazy to let both the alive and dead cat live in parallel universes after the measurement, and we are only in one of them and another universe has another version of you!

Further reading

(1) To learn more, the best way is to read the first a few sections of a proper quantum mechanics book. For example, Griffiths, *Introduction to Quantum Mechanics*, or Allan

Adams, Matthew Evans, and Barton Zwiebach's MIT Open Course on Quantum Mechanics (with videos).

(2) *The Principles of Quantum Mechanics* by Dirac, though a bit old, is still an excellent introduction.

(3) Some other famous books include Shanka, *Principles of Quantum Mechanics*, Sakuri, *Modern Quantum Mechanics*.

(4) For the developments of quantum mechanics, there is a popular science book in Chinese: Cao, *Does God Play Dice*.

What happens next in a university physics program?

(1) Quantum mechanics and advanced quantum mechanics. What I have covered is really a starting point and you need to learn a proper course. You need to learn the algebra of matrices, the quantum harmonic oscillator, angular momentum, approximation methods, and so on.

(2) Atom physics, solid state physics and material science. The explanation of matter is all based on quantum mechanics.

(3) Quantum field theory. This is how quantum mechanics works with special relativity.

Exercises

E4.1 Compton effect

Derive the wavelength shift (4.2) for the Compton effect.

E4.2 Time evolution of the eigenstates

What is the form of position eigenstates (quantum states with definite position) and momentum eigenstates for a freely moving particle? At the next moment in time (without measurement or interaction), are they still position and momentum eigenstates?

E4.3 Uncertainty of the Gaussian wave packet

Compute σ_x and σ_p of the Gaussian wave packet (4.23). Show that the uncertainty principle is satisfied in a saturated way.

E4.4 Commutation relations

If $\psi_q(x)$ is an eigenstate of x with eigenvalue q. Can $\psi_q(x)$ also an eigenstate of $\hat{p}$? Let us answer this question in different ways:

(1) Physically: Show that if $\psi_q(x)$ is at the same time eigenstate of x and $\hat{p}$, then the uncertainity principle is violated.

(2) Mathematically: Show that $\hat{p}(x\psi_q(x)) \neq x(\hat{p}\psi_q(x))$. Use this observation to prove that x and $\hat{p}$ do not share eigenstates.

E4.5 Time–dependent scattering on a potential barrier

Consider a particle moving towards a barrier-shaped potential (Fig. 4.26), and the incoming energy is lower than the height of the barrier (i.e. the potential energy on the barrier is higher than the total energy of the particle).

(1) Classically, can the particle pass to the other side of the barrier? Why?

(2) Quantum mechanically, can the particle pass to the other side of the barrier? Why? Explain it both in terms of continuity of the wave function, and energy conservation.

(3) Some ripples are observed in the figure below (snapshots of the probability density of the particle). Explain the ripples qualitatively.

E4.6 Scattering on a potential barrier: math and interpretations

(1) Derive the transmission amplitude E using A for a potential barrier.

(2) Show that the relations between A, B and E conserves probability.

(3) Show that E is exponentially suppressed if $E < V_2$. Find the exponential factor.

E4.7 Bound states

Consider bound states in a square potential well.

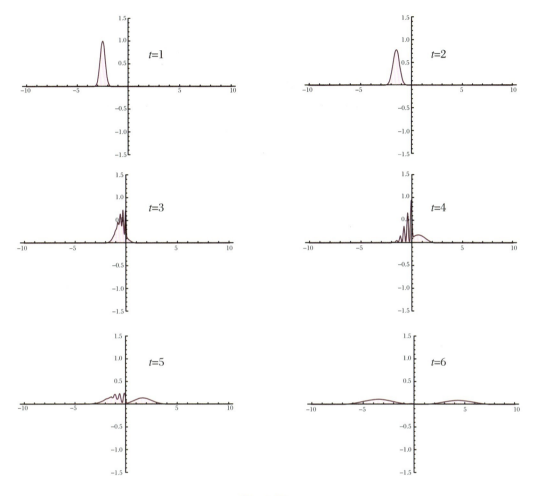

Fig. 4.26

(1) Count the number of free parameters and the number of constraint conditions, to show that energy needs to be quantized for bound states.

(2) When measuring the energy of a superposition of bound states with different energies, must the outcome of the measured energy be quantized?

(3) Keep the depth of the potential well fixed and increase the width. Shall the number of bound states increase or decrease? Why?

Chapter 5
Atoms

Feynman's question

As the opening of his lectures, Feynman asked the following question:

If, in some cataclysm, all of scientific knowledge were to be destroyed, and only one sentence passed on to the next generations of creatures, what statement would contain the most information in the fewest words?

He gave an answer himself:

Matter is made of atoms.

This statement is not absolutely right. For example, E&M waves may be considered a form of matter which is not made of atoms (though made of quanta). Dark matter is not made of conventional atoms either, and dark energy does not look like atoms by all means. Nevertheless, our familiar matter world is indeed made of atoms. And I agree that this is the message that we should pass on.

How do we know, why do we care, and what are the consequences that the world is made of atoms[①]? This will be the focus of this part. More explicitly, we will address:

① How did the atomic theory arise in chemistry?

② How do we know the size of an atom?

[①] With modern technology, ①, ② and ③ seems trivial. Because scanning tunnel microscopes can directly see and manipulate atoms. However, back to 150-200 years ago, how were these features known from scientific methods? Strictly speaking, they are not part of modern physics. But as it is not completely covered in general physics, I decide to include it here.

③ Can we find direct evidences for the existence of atoms?
④ How can an atom be stable? — How does quantum mechanics save the world?
⑤ Where do the chemical natures of atoms arise?

5.1 How Did We Know That Matter Is Made of Atoms?

The atomic theory was proposed by the ancient Greece (and other cultures) over 2000 years ago. However, at that time the atomic theory was based on philosophical guess and had little scientific support. The scientific atomic theory started from the 19th century.

Dalton's law of multiple proportions

Dalton (1808) noted that carbon can combine with oxygen in two different ways to form oxides. For fixed amount of carbon, the amount of oxygen needed in these two oxides is 1:2 — a ratio constitute of two small integers.

Is this observation a coincidence? Dalton tested other pairs of elements (which can form at least two different compounds). Similar properties are noted: Fixing the amount of one element, the amount of the other element needed for different compounds are ratios with small integers.

> **Is multiple proportion surprising?**
>
> One key talent as a physicist is to find out what scientific facts are ordinary, and what are surprising, which usually lead to breakthroughs in theory. Now the multiple proportion law is natural since we all know matter is made of atoms. But at Dolton's time it's surprising. Suppose that we don't know that matter is made of atoms. Then combining caron and oxygen toghether looks similar to spreading butter on bread — given one piece of bread, the amount of butter can continuously change. But Donlton's discovery is to say — there are two kinds of butter breads only, one consumes 5 g of butter and the other consumes 10 g per bread. No bread with 6 g butter can exist — isn't it surprising at that time?
>
> You may find the discussion here familiar — it looks similar to the discussion of photons for photoelectric effect — though here it's more trivial. In theory, it's equally surprising

that matter and photon are made of quanta, and it's equally surprising that they have wave properties. Just in practice some properties are easier to observe than others.

Such an observation deserves an explanation. Dalton pointed out that atomic theory could explain it very well. For example, in the carbon-oxygen example, one compound is CO and the other is CO_2. Thus obviously the amount of oxygen needed is 1 : 2. On the other hand, if there do not exist fundamental building blocks of matter, then the ratio wouldn't be always a simple ratio. This is one of the first scientific evidences of the atomic theory.

Oil film method

Independently, the size of atoms (molecules) can also be determined by oil film method. Franklin (1757) noted that oil can spread on a huge area of water. The thin film of oil upon water can be as thin as a single layer of molecule. However, such huge area is hard to measure. Is it possible to make the amount of oil smaller?

Langmuir (1917) used alcohol to dissolve oleic acad. Drip one drop of such solution to water. Alcohol is dissolved with water and oleic acad spread on the surface of water with an area measurable in a lab.

Now that there exist atoms, what are their natures? The kinetic theory of gases is developed to explain heat. Regarding the atomic theory, the next questions are: (1) how fast do the atoms move? and (2) what's the size of the atoms?

In the remainder of this section, we will focus on order-of-magnitude estimates without being careful about the coefficients. Particularly, when we use "$\sim$" in an equation, we have dropped the order one coefficients. We will also assume room temperature and 1 atm pressure. Also here we will not pay attention to the differences between atoms and molecules.

How fast do atoms move in gas?

In 1654, the Magdeburg hemisphere experiment demonstrated that air has huge pressure — 16 horses were needed to separate two hemispheres with the diameter of 50 cm, with vacuum in between them. The air pressure is 1 atm $\sim 10^5$ Pa. This is hard to imagine at that time since we are living in such huge pressure without knowing it; while easy to imagine now since there is kilometers of air above us, and they gravitate.

In the kinetic theory of gas, pressure is explained by atoms knocking each other (or the container). For ideal gas

$$P = \frac{1}{3} nmv^2 \tag{5.1}$$

where m is the mass of a single atom, n is number density of atoms, and v is the root mean square speed. For order-of-magnitude analysis, we consider v to be a typical speed of atoms' random motion.

Note that the density of gas is

$$\rho = nm \sim 1 \text{ kg/m}^3 \tag{5.2}$$

we have

$$v = \sqrt{3P/\rho} \sim 10^3 \text{ m/s} \tag{5.3}$$

Atoms move really fast[1]! They have to move fast against the pressure.

Do atoms move freely in gas?

Not always. They can collide. What is a typical distance L that an atom can move without collision? Technically, L is known as the mean free path[2].

We model atoms with hard balls with finite and fixed diameter d. The mean free path is thus

$$L \sim 1/(d^2 n) \tag{5.4}$$

Is it possible to measure L from simple experiments (without having access to the length as small as atoms)? A way to measure L is proposed by Maxwell (1859) through the viscosity of gas (Fig. 5.1). Here we provide a simplified pedagogical argument.

The viscosity μ of the gas is defined as

$$F_\parallel = \mu A u / y \tag{5.5}$$

---

[1] Alternatively, the same speed can be obtained from the specific heat of air $C_V \sim k_B/m \sim 1000$ J·kg/K^{-1} by noting that $k_B T \sim \frac{1}{2} mv^2$.

[2] L can be estimated as follows: to squeeze a box (of atoms) of length L and area A to one layer of atoms, the layer can fill a large part of area A. The number of atoms in the box is $N = nLA$. To squeeze the atoms to one layer, the total cross section of the atoms is $Nd^2 \sim A$. Thus $L \sim 1/(d^2 n)$.

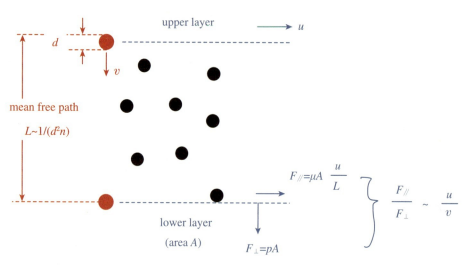

Fig. 5.1

where A is the area of each layer and y is the distance between layers. And u is the non-random part of the motion speed between layers. The viscosity of air is $\mu \sim 10^{-5} \text{kg} \cdot \text{m} \cdot \text{s}^{-1}$.

As shown in the above figure, we consider two layers of gas, separated by $y = L$. Then on average, the force is passed on from the upper layer to the lower layer by order-one collision. We expect that the force normal to the lower layer $F_\perp = pA$ can be related to $F_\parallel$ by

$$\frac{F_\parallel}{F_\perp} \sim \frac{u}{v} \tag{5.6}$$

since they come from the same collisions (having the same time duration of collision Δt in the impulse).

Putting the above equations together, the mean free path of air is thus

$$L \sim \frac{\mu v}{p} \sim 10^{-7} \text{ m} \tag{5.7}$$

Amazingly, without being able to see a single atom (at that time), one can infer the speed and mean free path of atoms! Further, can we also determine the size of atoms?

The size of an atom

Loschmidt (1865) determined the size of atoms with a very simple observation (based on the above mean free path). When gas liquify at low temperature, its volume condenses by a factor of about 1000 (and when liquid solidify the volume does not change much).

Chapter 5 Atoms 147

The number density of liquid and solid $\tilde{n}$ can be estimated by

$$\tilde{n} \sim d^{-3} \sim 1000n \sim 1000 d^{-2} L^{-1} \tag{5.8}$$

Here it is assumed that in liquid and solid the atoms are close to each other. Thus, the size of an atom is estimated by

$$d \sim 10^{-3} L \sim 10^{-10} \text{ m} \tag{5.9}$$

By the end of the 19th century, the atomic theory has gained great success. As Sir William Thomson commented in 1889, the atomic theory is

> ... founded respectively on the undulatory theory of light, on the phenomena of contact electricity, on capillary attraction, and on the kinetic theory of gases, agrees in showing that the atoms or molecules of ordinary matter must be something like the 1/10,000,000th, or from the 1/10,000,000th to the 1/100,000,000th, of a centimeter in diameter.

However, the atomic theory still receive some criticism (for example, by Mach) since no one has actually seen an atom. Is the atomic theory just an "effective" theory to explain experiments, or are atoms indeed the building blocks of our *real* world?

(Optional) Brownian motion

In 1827, Robert Brown noticed that tinny particles from pollen "dance" in the water (Fig. 5.2). Einstein in 1905 explained the dancing particles by the motion of molecules — we actually see their direct effects. Here we use Langevin's stochastic differential equations (1908) to describe Einstein's idea.

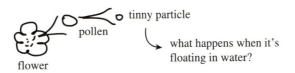

Robert Brown (1827) observed it under microscope:

Fig. 5.2

The particle from pollen is constantly kicked by water molecules. The speed of the particle satisfy the following Langevin equation

$$M\frac{dv}{dt} = -\lambda v + f(t) \tag{5.10}$$

where

(1) The parameter M is the mass of the particle.

(2) The parameter λ describe a friction force, i.e. dissipation in water. This dissipation can be measured by, say, measuring the terminal velocity of a heavy ball in deep water.

(3) The function $f(t)$ is force from random kick on the particle by molecules. The random force can be modeled by white noise[①]:

$$\langle f(t)\rangle = 0, \qquad \langle f(t)f(t')\rangle = \Lambda\delta(t-t') \tag{5.11}$$

Similarly to the notation in quantum mechanics, here $\langle\cdots\rangle$ means taking average. Λ can be measured as follows: given a set of particles initially at coincide positions, due to the fluctuation, the particles diffuse and one can obtain Λ by the speed of diffusion for the set of particles. The Langevin equation (5.10) can be solved by "variation of constants" method:

$$v(t) = v_0 e^{-\frac{\lambda}{M}t} + \frac{1}{M}\int_0^t dt_1 f(t_1) e^{-\frac{\lambda}{M}(t-t_1)} \tag{5.12}$$

Thus,

$$\langle v(t)\rangle = \langle v_0\rangle e^{-\frac{\lambda}{M}t} \tag{5.13}$$

$$\langle v^2(t)\rangle = \langle v_0^2\rangle e^{-2\frac{\lambda}{M}t} + \frac{1}{M^2}\int_0^t dt_1 \int_0^t dt_2 e^{-\frac{\lambda}{M}(t-t_1)} e^{-\frac{\lambda}{M}(t-t_2)} \langle f(t_1)f(t_2)\rangle$$

$$= \langle v_0^2\rangle e^{-2\frac{\lambda}{M}t} + \frac{\Lambda}{2M\lambda}\left(1 - e^{-2\frac{\lambda}{M}t}\right) \tag{5.14}$$

After a long time t, the exponentially suppressed terms (memory about the initial condition) dies out. As a result, we have vanishing $\langle v(t)\rangle$, and[②]

$$\frac{1}{2}M\langle v^2(t)\rangle = \frac{\Lambda}{4\lambda} \tag{5.15}$$

We have thus computed the average kinetic energy of the particle. Note that this energy should be statistically the same as the kinetic energy of a molecule in the water. Because

---

[①] To very good approximation, the random kicks are independent of each other. Thus the kicks at different time has no correlation between each other. But at the same time, the "same" kick has self correlation. This is why $\langle f(t)f(t')\rangle$ is propotional to $\delta(t-t')$. The parameter Λ characterize how fast the particle fluctuates.

[②] Note that $\frac{1}{2}M\langle v^2(t)\rangle = \frac{1}{2}k_\mathrm{B}T$. Thus this is a direct measurement of the Boltzmann's constant k_B. In 1908, such a measurement is made by Perrin.

otherwise the particle will transfer energy to water molecule and lose energy (equipartition theorem). Compared to total internal energy per unit volume, we know how many atoms there are and thus their size.

> **Distance ladders**
>
> We encountered three methods of measuring the size of an atom. In each case, a distance ladder is built from human accessible scales to a microscopic scale.
>
> Oil film: through oil-alcohol solution.
>
> Kinetic theory: through three large or small numbers-air pressure, viscosity and condensation rate from gas to liquid.
>
> Brownian motion: through motion of tinny particle from pollen.
>
> Distance ladder is an important idea for accessing distances that we are not yet able to reach. This is true also for longer distances, for example the measurement of distance in astronomy and cosmology.
>
>

(Optional) Fluctuation and dissipation: another equivalence from Einstein

The interpretation of Brownian motion is another Einstein-style discovery. Let's take a closer look at the two measurable quantities involved in Brownian motion: Λ denotes the strength of fluctuation; and λ denotes the rate of dissipation. They are at the first sight unrelated. But Einstein noted that they are of the same origin.

Why there is dissipation? Macroscopically it is from viscosity or friction. While microscopically it is from the fact that, imagine that you are the particle, once you run forward, there are more molecules hitting your face than your back, with a greater speed. Thus the random knocks tend to block your motion. This related dissipation to fluctuation.

This observation can be generalized to a fluctuation-dissipation theorem in statistical physics. It has a wide range of applications. For example, the thermal noise (Johnson-Nyquist noise) in a resistor is related to its resistance by the same observation.

5.2 The Hydrogen Atom

Now that we have understood that matter is made of atoms, let us study the nature of atoms. We start from the simplest atom: the hydrogen atom.

5.2.1 Aspects of Observations

Since Newton, lots of scientists studied the spectrum of light (Fig. 5.3). Since matter is made of atoms, light emitted/absorbed by matter is also (usually) emitted/absorbed by atoms as well. Can we learn about the nature of atoms from the study of light spectrum?

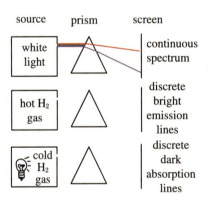

Fig. 5.3

Developments in spectroscopy

(1) 1666, Newton: using prism to split the sunlight into a spectrum.

(2) 1814, Fraunhofer: dark lines in the sun spectrum.

(3) 1817—1823, Fraunhofer: spectra from the moon, Venus and Mars has some dark lines at the same position of the sun spectrum, and also some new lines.

(4) 1826, Herschel and Talbot: emission spectrum, different elements have different spectra (spectra to elements like fingerprints to human).

(5) 1832, Brewster: dark lines in the sun spectrum are absorption lines by its atmosphere.

(6) 1859, Kirchhoff: emission and absorption lines have the same position.

(7) 1860—1861, Kirchhoff and Bunsen: discovered elements (cesium and rubidium) by their spectra (like catching criminals by fingerprints).

(8) 1885, Balmer: found a formula for the Hydrogen spectrum for visiable light frequencies, see Fig. 5.4. In the below equation, R_H is a constant.

$$\frac{1}{\lambda} = R_H \left(\frac{1}{4} - \frac{1}{n^2} \right), \quad n = 3, 4, 5, \ldots \tag{5.16}$$

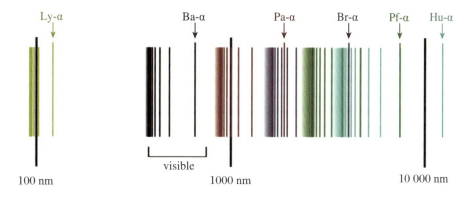

Fig. 5.4 The spectral lines of Hydrogen atom

(9) 1888, Rydberg: proposed a general formula for Hydrogen

$$\frac{1}{\lambda} = R_H \left(\frac{1}{m^2} - \frac{1}{n^2} \right), \quad n = m+1, m+2, \cdots \tag{5.17}$$

(10) 1906, Lyman: ultraviolet bands of Hydrogen spectrum ($m = 1$ of (5.17)).

(11) 1908, Paschen: infrared bands of Hydrogen spectrum ($m = 3$ of (5.17)).

How to understand such spectra of elements? A simple equation such as (5.17) deserves a theory. We need to understand in details how atom interacts with light — knowing the inner structure of atoms should help. Thus, let us review briefly another line of research in the history — the subatomic structure.

The discovery of the subatomic structure

It was shocking to know that atoms are actually not the most fundamental building blocks of matter. They are made of even smaller particles.

(1) 1897, Thomson: discovery of electron.

(2) 1899, Rutherford: discovery of α particle (Helium nuclei).

(3) 1904, Thomson: the plum pudding model of atom.

(4) 1909, Geiger, Marsden, Rutherford, α particle scattering.

(5) 1911, Rutherford: nuclei-electrons model, like sun-planets that electrons run in orbits around the nuclei. The nuclei radius is about 10^{-5} of an atom.

Rutherford's atom model has a serious problem: Electron must have acceleration when moving within the atom volume. However, acceleration of charge implies E&M radiation. The electron should then lose energy and fall onto the nuclei.

How to stabilize Rutherford's atom and save the world?

Now, both the studies of spectroscopy and subatomic structure have evolved to the 1910s. At that time physicists started to appreciate the need of quantum mechanics. Would quantum mechanics play a role to solve the problem of subatomic structure, and/or explain the atom spectrum?

5.2.2 Bohr's Model of the Hydrogen Atom

Bohr's model

In 1913, Bohr brought together 3 aspects of physics, to build up his atomic model:

(1) Atomic spectroscopy. There are different frequencies of light:

$$\nu^{(n_2,n_1)} = \frac{c}{\lambda^{(n_2,n_1)}} = cR_H \left(\frac{1}{n_2^2} - \frac{1}{n_1^2} \right) \tag{5.18}$$

(2) Quantum nature of photon. Frequency of light originates from the energy of photon: $E_\gamma^{(n_2,n_1)} = h\nu^{(n_2,n_1)}$.

(3) Subatomic structure. In a hydrogen atom, the electron can be on different orbits. Classically, those orbits correspond to different energy of the electron. First note force:

$$F = \frac{mv^2}{r} = \frac{1}{4\pi\epsilon_0} \frac{e^2}{r^2} \quad \Rightarrow \quad mv^2 = \frac{1}{4\pi\epsilon_0} \frac{e^2}{r} \tag{5.19}$$

Thus the energy of the electron is

$$E_e = \frac{1}{2}mv^2 + V(r) = -\frac{1}{2}mv^2 \tag{5.20}$$

Now let's put them together. The frequency of the electron on such an orbital is $\nu_e = v/(2\pi r)$. It is natural to expect that for an electron to move from infinity at rest to a orbital, the frequency of photons emitted is related to this frequency. Bohr suggests that[1]

---

[1] Here the factor of 2 is a guess, probably as a naive average between ν_e and 0, which is the inital frequency for an electron at infinity. Don't worry, we will soon get rid of Bohr's atom. The Schrödinger equation can compute everything clearly.

$$\nu = \frac{\nu_e}{2} = \frac{v}{4\pi r} \tag{5.21}$$

As the emission is quantized, we have

$$|E_e| = nh\nu \quad \Rightarrow \quad mvr = \frac{nh}{2\pi} \tag{5.22}$$

where n is an integer. Using (5.19) to cancel r, (5.22) can be written as

$$v = \frac{e^2}{2n\epsilon_0 h} \quad \Rightarrow \quad E_e(n) = -\frac{me^4}{8\epsilon_0^2 h^2} \times \frac{1}{n^2} \tag{5.23}$$

This indeed derives Eq. (5.18):

$$\nu^{(n_2, n_1)} = cR_\mathrm{H}\left(\frac{1}{n_2^2} - \frac{1}{n_1^2}\right), \quad R_\mathrm{H} = \frac{me^4}{8\epsilon_0^2 h^3 c} \tag{5.24}$$

Relation to de Broglie's matter wave

The quantization condition (5.22) can be understood in an intuitive way: the equation can be rewritten as

$$\frac{p}{\hbar} = \frac{n}{r} \tag{5.25}$$

where the momentum of the electron is $p = mv$, and $\hbar = h/(2\pi)$.

Requiring the electron on the orbital being standing wave (satisfying periodic boundary condition), we note that $n/r = k$ is the wave number. And thus Eq. (5.22) is nothing but the de Broglie's matter wave $p = \hbar k$ (which is proposed in 1924, historically much later than Bohr's atom, Fig. 5.5).

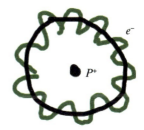

Fig. 5.5

Although Bohr's theory of atom is successful in explaining the Hydrogen spectrum to certain accuracy, it is unable to explain the fine structure of Hydrogen, or the spectra of multi-electron atoms.

Now we know that those failures of Bohr's model is because of the semi-classical nature of Bohr's model. The electron's position and momentum satisfies the uncertainty principle. Thus there is no classical orbits. Those difficulties are indeed gone after the full quantum mechanics is taken into account.

5.2.3 (Optional) The Schrödinger Equation of the Hydrogen Atom

In this subsection, we present the technical details for solving the Schrödinger equation, as optional material. We will scratch the procedure without presenting the full details here.

The 3-dimensional Schrödinger equation

In 3 dimensions, the stationary state Schrödinger equation is[1]

$$\left[-\frac{\hbar^2}{2M}(\partial_x^2+\partial_y^2+\partial_z^2)+V\right]\psi = E\psi, \qquad V \equiv V(r) \propto \frac{1}{r} \qquad (5.26)$$

or equivalently,

$$(\partial_x^2+\partial_y^2+\partial_z^2)\psi + \frac{2M}{\hbar^2}(E-V)\psi = 0 \qquad (5.27)$$

Now we get a Schrödinger equation with derivatives in all x, y, z directions. A differential equation with different types of partial derivatives is known as a partial differential equation. How to solve it?

Spherical coordinates

To solve this equation, one first note that the potential has spherical symmetry. Thus spherical coordinates fits better to solve the equation. In spherical coordinates, the differential operators of the above equation is rewritten into

$$(\partial_x^2+\partial_y^2+\partial_z^2)\psi = \frac{1}{r^2}\partial_r(r^2\partial_r\psi) + \frac{1}{r^2\sin\theta}\partial_\theta(\sin\theta\partial_\theta\psi) + \frac{1}{r^2\sin^2\theta}\partial_\phi^2\psi \qquad (5.28)$$

Spherical harmonics

In fact, we can pack a good part of the procedure here into known math — the solution of the angular directions can be directly written as the spherical harmonics: $\Theta(\theta)\Phi(\phi) \propto Y_{\ell m}(\theta,\phi)$. The spherical harmonics $Y_{\ell m}(\theta,\phi)$ is already well studied in the 18—19 century for studying classical wave equations on a sphere. But here we still use $\Theta(\theta)\Phi(\phi)$ since you may not be familiar to $Y_{\ell m}(\theta,\phi)$.

---

[1] In this section, we denote the mass of the quantum particle by M instead of m because traditionally m is used to label the magnetic quantum number (see definition below).

Separation of variables

Now we use a smart trick — separation of variables. This is a powerful method in solving a class of partial differential equations. We assume that the solution takes the form

$$\psi(r,\theta,\phi) = R(r)\Theta(\theta)\Phi(\phi) \tag{5.29}$$

The Schrödinger equation can now be written as

$$\frac{1}{R}\partial_r\left(r^2\partial_r R\right) + r^2\frac{2M}{\hbar^2}[E - V(r)] + \frac{1}{\Theta\sin\theta}\partial_\theta(\sin\theta\partial_\theta\Theta) + \frac{1}{\Phi\sin^2\theta}\partial_\phi^2\Phi = 0 \tag{5.30}$$

This equation has an interesting structure: the first line of the equation only depends on r and the second line of the equation does not depend on r. As a result, the first line cannot depend on r either and thus must be a constant:

$$\frac{1}{R}\partial_r\left(r^2\partial_r R\right) + r^2\frac{2M}{\hbar^2}[E - V(r)] = \ell(\ell+1) \tag{5.31}$$

With this observation, the second line of (5.30) can be written as

$$\frac{\sin\theta}{\Theta}\partial_\theta(\sin\theta\partial_\theta\Theta) + \ell(\ell+1)\sin^2\theta + \frac{1}{\Phi}\partial_\phi^2\Phi = 0 \tag{5.32}$$

Again, this new equation has similar structure: The first line depends only on θ and the second line does not depend on θ. Thus both of them must be constants. We thus have

$$\frac{\sin\theta}{\Theta}\partial_\theta(\sin\theta\partial_\theta\Theta) + \ell(\ell+1)\sin^2\theta = m^2 \tag{5.33}$$

$$\partial_\phi^2\Phi + m^2\Phi = 0 \tag{5.34}$$

The partial differential equation, outstandingly, factorize into three ordinary differential equations: Eqs. (5.31), (5.33) and (5.34). Ordinary differential equations are much easier to deal with. Nevertheless, we are not going to fully solve those equations for you. We are going to solve Eq.(5.34) only as it is a familiar wave equation. We give the physical result of Eq. (5.31) and Eq. (5.33) without diving into full details.

The Φ equation: the magnetic quantum number m with $L_z = m\hbar$

It is easy to check that Eq. (5.34) has solution[1]

$$\Phi = e^{im\phi} \tag{5.35}$$

---

[1] In general, we have $\Phi = ce^{im\phi}$ with a constant c. This constant is not relevant and is omitted. Also, there is another set of solution $\Phi = e^{-im\phi}$. That solution corresponds to $m \to -m$.

They are just left and right moving waves in the angular direction (wrt the z-axis). Physically, m is related to the angular momentum. The angular momentum in quantum mechanics is $\boldsymbol{L} = \boldsymbol{r} \times \boldsymbol{p} = -i\hbar(r \times \nabla)$. One can derive that $L_z = -i\hbar \partial_\phi$ in spherical coordinates. As a result, $L_z = m\hbar$.

Note that the wave has periodic boundary condition (recall the continuity requirements of the wave function):

$$\Phi|_{\phi=0} = \Phi|_{\phi=2\pi}, \qquad \partial_\phi \Phi|_{\phi=0} = \partial_\phi \Phi|_{\phi=2\pi} \tag{5.36}$$

To satisfy those conditions, m must be integers and is known as the magnetic quantum number[①].

The Θ equation: the azimuthal quantum number ℓ

Now we look at Eq. (5.33). We shall not solve the equation here (you will learn the solution in a proper QM class). But recall our experience of the ϕ direction: There are two similarities:

(1) Note that θ is also rotation so should be related to some sort of angular momentum. As a result, ℓ should be related to some sort of angular momentum. As we will see soon, the total momentum of the wave function is

$$L^2 = \ell(\ell+1)\hbar^2 \tag{5.37}$$

(2) For the solution $\Theta(\theta)$ to be regular at $\theta = \pi$, it turns out that ℓ should take integer values only. It turns out that ℓ (known as the azimuthal quantum number) can only take integer values starting from m: $\ell = m, m+1, \ldots$ We can rearrange this requirement into

$$\ell = 0, 1, 2, 3, 4, 5, 6, \ldots \tag{5.38}$$

and those orbitals are known as s, p, d, f, g, h, i, ... in chemistry. For a given ℓ, m is restricted to be[②]

$$m = 0, \pm 1, \ldots, \pm \ell \tag{5.39}$$

[①] The math structure here is similar to Bohr's semi-clssical atom. However, the physics is different. We did not confine the electron to move along the ϕ-direction only. Rather, now we are just calculating the wave function on the ϕ-direction. The electron has probability to appear anywhere instead only on the orbital.

[②] As we have argued that m is related to the angular momentum in the z-direction and ℓ is related to the total angular momentum, it is natural to understand that m should not exceed ℓ.

> **The total angular momentum**
>
> Adding a factor $\frac{\hbar^2}{2Mr^2}\ell(\ell+1)$ to the potential looks familiar. In classical mechanics, when reducing a 3D central-force problem into 1D, the additional term added into the effective potential is $\frac{L^2}{2Mr^2}$, where L is the total angular momentum of the electron. Comparing it to (5.40), we conclude that the angular momentum of the system is $L^2 = \ell(\ell+1)\hbar^2$.

The R equation: The principal quantum number n

Finally, we will get some feeling from Eq. (5.31), again without fully solving it. We can rewrite Eq. (5.31) as

$$\frac{1}{R}\partial_r\left(r^2\partial_r R\right) + r^2 \frac{2M}{\hbar^2}\left[\left(E - \frac{\hbar^2}{2Mr^2}\ell(\ell+1)\right) - V(r)\right] = 0 \tag{5.40}$$

This is similar to a stationary Schrödinger equation, with reduced energy. The reduction of energy is because we are interested in a reduced one-dimensional problem and thus must deduct the energy coming from rotation.

This is a bound state problem. Thus the energy spectrum is discrete. After solving Eq. (5.40) (which is technical and we skip the details here), one eventually finds

$$E = \frac{E_1}{n^2}, \quad E_1 \equiv -\frac{me^4}{32\pi^2\epsilon_0^2\hbar^2} \tag{5.41}$$

Here n is an integer with $n = \ell+1, \ell+2, \ldots$ This is the result of quantized energy levels that Bohr obtained.

It is intuitive that the starting integer for n must be bounded by ℓ because the larger ℓ, the larger energy from angular momentum, to be deducted from E. One can again rewrite the requirements for n, ℓ, m as:

$$n = 1, 2, \ldots, \quad \ell = 0, 1, \ldots, n-1, \quad m = 0, \pm 1, \pm 2, \ldots \pm \ell \tag{5.42}$$

5.2.4 Properties of the Hydrogen Atom

As the previous subsection is optional and you may like to skip that, here we summarize the properties of the atom, and discuss their physical consequences.

Number of possible atomic states

Given n, ℓ, m, we can count the number of possible states. This is of crucial importance in chemistry. For the counting, we first note two things about electrons:

- Electrons are fermions and thus they cannot be in the same state.

- For each given set of n, ℓ, m, there are two electron states. This is because electron has spin $m_e = \pm 1/2$, namely spin up and spin down.

We can now count the number of possible atomic states. For example, for $n = 1$, we must have $\ell = 0, m = 0$ thus two states (because of spin of electron). For $n = 2$, we have

$$\begin{cases} \ell = 0, m = 0, m_e = \pm 1/2, & \text{2 states} \\ \ell = 1, m = 0, m_e = \pm 1/2, & \text{2 states} \\ \ell = 1, m = \pm 1, m_e = \pm 1/2 & \text{4 states} \end{cases} \quad (5.43)$$

thus in total 8 states. In general, for given n, there are $2n^2$ states.

The states with the same n is said to be in the same shell; the states with the same n and ℓ is said to be in the same orbital.

The electron cloud

Now with the wave function $\Psi(r, \theta, \phi)$, the property of the electron is expressed in terms of the probability density amplitude (such a distribution is known as the electron cloud). This is different from classical orbitals. Nevertheless, we can still use the term "orbitals" to denote the states with different n, ℓ and m. Some of the states are illustrated in Fig. 5.6.

For each n, the s, p, d, ... orbitals has the features illustrated below. One notes that the s orbital has greater probability density at both small r and large r than p and d.

Pay attention to the oscillations in the probability distribution (Fig. 5.7). Although we have not calculated them explicitly here, they are understandable from our experience in the bound states in an infinitely deep potential well.

Transitions and selection rules

Now that we know the Hydrogen atom (in general all atoms) can be in stationary states with definite energy. There can be transition between those states. Namely, an

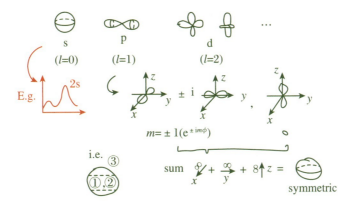

Fig. 5.6

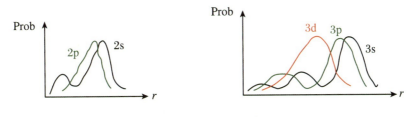

Fig. 5.7

atom can emit a photon and become a lower energy state; or absorb a photon and become a higher energy state. From energy conservation, the frequency of the photon is

$$\nu = \frac{E_{n'} - E_n}{h} \tag{5.44}$$

where the two states $E_{n'}$ and E_n must satisfy $\Delta \ell = \pm 1$, $\Delta m_\ell = 0, \pm 1$. This is because a photon has spin 1.

5.3 The Periodic Table

Now we are interested in elements other than Hydrogen.[1] What happens for an atom with a nuclei with Z positive charges?

---

[1] For multiple electron atoms — even for helium, the Schrödinger equation cannot be solved analytically. One may either try to solve it numerically or study the qualitative features from our experience of the hydrogen atom.

Qualitative differences for many-electron atoms

(1) Pauli's exclusion principle: If one orbital is filled, the other electrons in the atom must be put in other orbitals.

(2) Screening and other electron interactions: ① Classically, the inner electrons screen the charge of the nuclei when we consider the motion of outer electrons. In quantum mechanics, there is no absolute inner/outer electrons. But the spatial distribution of electrons needs to be considered in the screening effect. As a result, in a hydrogen atom for the same n, s, p, d, f, ..., orbitals has the same energy. But for multiple electron atoms, for the same n, $E_s < E_p < E_d < E_f$ because higher angular momentum means (on average) farther from the nuclei and thus more screening.

To see how these features affect the electron configurations of elements, especial the outermost occupied shell (called the valence shell which is mostly responsible for its chemical properties), we list some elements and how the electrons are configured in them. We will only discuss the ground states.

Electron configurations of some elements

(1) Hydrogen: For the ground state of hydrogen, the electron is in $n=0$ state (Fig.5.8). There is only a s-orbital ($\ell = 0$). The electron has an ionization energy (the minimal energy to remove an electron to infinity) of 13.6 eV.

(2) Helium: Naively, the ionization energy of helium would be $Z^2 \times 13.6$ eV $= 54.4$ eV. However, actually its ionization energy is 24.5 eV. This is because of screening. In fact, after taking away the first electron of helium, we indeed need about 54.4 eV to take its second electron away (the second ionization energy).

As helium has full filled its valence shell, it is hard to offer or accept electrons to other atoms to form chemical bound. Thus, its chemical property is very stable.

(3) Lithium: With 3 electrons, one electron has to go to the $n=2$ shell. Will the electron be filled in the 2s ($\ell = 0$) orbital or the 2p ($\ell = 1$) orbital? Due to screening, the s-orbital has lower energy thus the electron is filled in the s-orbital. The ionization energy of lithium is 5.45 eV. It's very easy to lose such an electron and thus lithium has active chemical property.

(4) From beryllium to Argon, we fill additional electrons as discussed above: 2s, 2p, 3s, 3p.

---

① In general electron interactions are very complicated. For example, in some sense the Helium electron interaction problem is a quantized 3-body problem. Fortunately, the "screening" point of view can provide us some intuitions and is sometimes a good approximation.

Chapter 5 Atoms 161

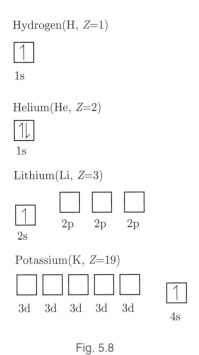

Fig. 5.8

(5) Potassium: Naively we would have proceeded to fill 3d after 3p. However, as the 3d orbital has too large angular momentum, it actually have slightly greater energy than 4s. As a result, for potassium, 4s is filled first.

Hopefully, the above discussion provides some physical understanding for the periodic table. (Fig. 5.9)

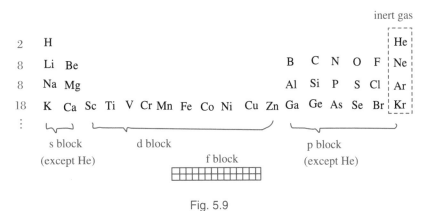

Fig. 5.9

5.4 Epilogue: Summary and What's Next

Further reading

(1) For the size of the atom, more details can be found in Maxwell and Loschmidt's original works. Maxwell's paper contains lots of technical calculations while Loschmidt's paper is very simple to understand.

(2) Although Bohr's atomic model is exceeded by the Schrödinger equation approach, if you are interested in more historical details, you can read Bohr's origonal paper.

(3) For the Schrödinger equation and the periodic table, more details can be found at the Feynman Lectures III Chapter 19. For even more details related to chemistry, you may like to read a book on physical chemistry, for example Physical Chemistry by Silbey, Alberty and Bawendi.

What happens next?

The quantum part of the content here will be studied in more details in quantum mechanics and physical chemistry.

The atomic and sub-atomic structure of matter is just a starting point of exploring the inner matter structures, instead of an end. The research on particle physics studies more elementary particles. We will encounter more details on particle physics and beyond in a later chapter, From Particles to Strings.

Exercises

E5.1 Bohr's atom from matter wave

Bohr's assumption $\nu = \nu_e/2$ seems mysterious. Without using this assumption, derive R_H using other assumptions in Bohr's atomic model, making use of the de Broglie's matter wave.

E5.2 Atomic states

For a hydrogen atom:

(1) How many electron states are there with principal quantum number $n=3$?

(2) Among those $n=3$ states, how many states are there in the s, p and d orbitals, respectively?

(3) Let the energy of this $n=3$ state be E_3. Now a photon is emitted due to transition evolving (i.e. $n=3$ being the initial or the final state) the $n=3$ state. What is the highest possible frequency of the photon, in terms of E_3 and the Planck constant h? (Note: You don't have to consider fine and hyper-fine structures of the atom.)

E5.3 Electron configurations

Find electron configurations for elements Be, C, Ne and K.

Chapter 6

Entanglement and Quantum Information

Alice's continued adventures in the quantum wonderland

Alice continues to find more about her electron friends:

① Superposition of a "standing" electron and a "handstand" electron is a "lied-down" electron (Fig. 6.1). The head direction of the lied-down electron depends on the coefficient of the superposition.

Fig. 6.1 ①

② Alice has very good memory. But she finds it a nightmare trying to remember the relations between a handful of electrons.

③ Alice can show that a measurement of an electron light years away can *immediately* affect the electron state around her. But this superluminal "quantum information" does not conflict the causality of special relativity.

④ Alice wants to find out the origin of randomness of the quantum world — is it due to any hidden local objects that she cannot see? Her electron friends are able to falsify

---

① Sorry, the above state does not look like an electron. Suppose it is an electron and we will make it clearer later.

this for her.

Two approaches to the same quantum mechanics

Quantum mechanics can be learned in two ways:

① By first studying the classically most familiar object. The previous chapter corresponds to this approach, focusing on the position and momentum of a particle.

② By first studying the simplest quantum object. In this part, we will introduce a feature "spin" of a quantum particle, with no classical counterpart.

Though these two types of systems appear very different, they follow the same quantum mechanical laws.

Why we use spin for entanglement?

In principle, we would use the familiar position and momentum for entanglement. However, theoretically, continous states are difficult to deal with. Operationally, position eigenstates spread and momentum eigenstates run away. Thus it is much easier (though still extremely difficult) to construct actual devices with spin about quantum entanglement, as the theoretical basis of quantum information and quantum computing.

6.1 Spin

6.1.1 The Stern-Gerlach Experiment

Before stepping into the quantum mechanical world again, let us review a concept which has classical counterpart: the magnetic moment.

The magnetic moment

An object with a magnetic dipole moment (or magnetic moment for short) $\boldsymbol{\mu}$ has a magnetic potential energy U when it is put in a magnetic field $\boldsymbol{B}$①

$$U = -\boldsymbol{\mu} \cdot \boldsymbol{B} \qquad (6.1)$$

In our classical world, we are familiar with magnetic moments, and the simplest example is a magnet. A magnet can be attracted (or repulsed) by another magnetic because of Eq. (6.1). Other examples of classical magnetic moment include rotating charges.

Do atoms have magnetic moment? The Stern-Gerlach experiment

Let us now *measure* the magnetic moment of atoms. Consider a beam of hot silver atoms passing through an *inhomogeneous* magnetic field, and the out-going silver atoms leave records on a screen (Fig. 6.2). (The condition "hot" indicates that the atoms in the beam have random orientations.)

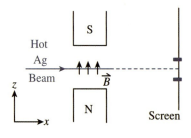

Fig. 6.2

If we use our *classical* experience to imagine the hot beam of silver atoms. What do we expect? We expect that the result on the screen would be one of the following:

(1) If silver atoms have no magnetic moment: we still have one beam of silver atoms; the external magnetic field has no effect on the beam.

(2) If silver atoms have nonzero classical magnetic moment: Imagine each silver atom is a magnet with random (and continuously variable) orientation. These magnets experience different forces depending on their orientations, and thus in the limit that there are many magnets, the beam will spread in a continous way.

What is the *actual* outcome of the Stern-Gerlach experiment?

One beam splits into two beams with equal strength.

---

① Thus, if the magnetic field is inhomogeneous, there is a force acting on the magnetic moment: $\boldsymbol{F} = -\partial U/\partial \boldsymbol{r}$.

The split of the beam indicates that a silver atom has a magnetic moment $\boldsymbol{\mu}$. However, why the split is discrete, instead of continuous?

The quantum mechanical interpretation of the Stern-Gerlach experiment

We observe two quantum mechanical aspects of the silver atom from the experiment:

(1) The magnitude $|\boldsymbol{\mu}|$ of the silver atom is quantized and cannot vary continuously.

(2) What about the direction $\hat{\boldsymbol{\mu}}$? We have two observations:

First, from rotational invariance of the laws of nature, and statistical property of the hot beam, we expect that *before measurement*, i.e. before coming to the external magnetic field, for each silver atom $\boldsymbol{\mu}$ should point to random directions.

However, the external magnetic field *measures* the z-component of the magnetic moment μ_z of the atoms, and thus the state of each atom collapses to one of the two eigenstates with eigenvalue $\pm|\boldsymbol{\mu}|$, respectively.

These explains the Stern-Gerlach experiment.

Let us now call the two eigenstates of the magnetic moment in the z-direction

$$|\uparrow\rangle \text{ with eigenvalue } |\boldsymbol{\mu}|, \text{ and } |\downarrow\rangle \text{ with eigenvalue } -|\boldsymbol{\mu}| \qquad (6.2)$$

The other orientations are superpositions of $|\uparrow\rangle$ and $|\downarrow\rangle$

What about states with other orientations, for example the states pointing inward $|\otimes\rangle$, outward $|\odot\rangle$, left $|\leftarrow\rangle$ and right $|\rightarrow\rangle$?

As discussed above, after measurement along the z-direction, they collapse into states $|\uparrow\rangle$ and $|\downarrow\rangle$. They thus should be superpositions of $|\uparrow\rangle$ and $|\downarrow\rangle$.

To further confirm this idea, one can string two Stern-Gerlach experiments in a row:

(1) The first magnetic field is in the y-direction.

(2) We put another magnet with magnetic field in the z-direction before observing the outcome on the screen.

The outcome of the experiment is: after passing the first magnet, the beams split into two with $|\otimes\rangle$ and $|\odot\rangle$ states; and each beam further split into $|\uparrow\rangle$ and $|\downarrow\rangle$ after passing the second magnet. Again, the split beams have equal strength.

From atoms to electrons

An electron itself also have magnetic moment. However, if we have used an electron in the Stern-Gerlach experiment, the dominate effect will be the Lorentz force and the magnetic moment is too small to observe. In fact, the magnetic moment of the silver atom comes from the magnetic moment of its outermost (5s) electron, thus indeed shows that the electron has magnetic moment.

Spin and angular momentum

Further experiments show that the magnetic moment of a quantum particle is related to its angular momentum. For an electron,

$$\boldsymbol{L} = \frac{1}{2} \frac{m_e}{(-e)} \boldsymbol{\mu} \tag{6.3}$$

where e and m_e are charge and mass of an electron, respectively.

For historical reasons, the intrinsic magnetic moment of an electron is called "spin". In fact, the term "spin" is more general than magnetic moment for fundamental particles. Because neutral particles like photons have intrinsic angular momentum as well, but no magnetic moment.

> **Nothing is spinning**
>
> The name "spin" seems misleading: it provides an intuition that the electron is spinning classically, and thus has a magnetic moment (a spinning charged ball indeed has a magnetic moment classically).
>
> However, for an electron, the spin is a fundamental and instric feature, instead of any rotational motion, because:
>
> No experiment shows the electron is actually rotating.
>
> If one uses a rotating ball to explain spin, the rotation speed on the surface of the ball is faster than light, which is not possible.

6.1.2 The Quantum Mechanics of Spins

In this subsection, we summarize the quantum properties of the spin system. This is a nice application of the quantum mechanics part, and a foundation for studying entanglement.

Basis and some other spin states

In the quantum mechanical language, the indication of equal strength beams in a sequence of Stern-Gerlach experiments is[①]

$$|\rightarrow,\leftarrow,\otimes,\odot\rangle = \frac{1}{\sqrt{2}}\left(|\uparrow\rangle + e^{i\phi}|\downarrow\rangle\right) \tag{6.4}$$

The phase turns out to indicate the freedom to choose y and z axes when the x axis is specified. It is convenient to choose:

$$|\rightarrow\rangle = \frac{1}{\sqrt{2}}(|\uparrow\rangle + |\downarrow\rangle), \qquad |\leftarrow\rangle = \frac{1}{\sqrt{2}}(|\uparrow\rangle - |\downarrow\rangle) \tag{6.5}$$

$$|\otimes\rangle = \frac{1}{\sqrt{2}}(|\uparrow\rangle + i|\downarrow\rangle), \qquad |\odot\rangle = \frac{1}{\sqrt{2}}(|\uparrow\rangle - i|\downarrow\rangle) \tag{6.6}$$

A qubit of information

A spin state can be thought as the building blocks of information — a quantum bit (qubit for short) of information. This is because the state has two basis $|\uparrow\rangle$ and $|\downarrow\rangle$. This is to be compared to that a state taking 0 or 1 contains one classical bit of information.

How much information is contained in a qubit?

We need to generalize Eq. (6.4), since in Eq. (6.4), $|\uparrow\rangle$ and $|\downarrow\rangle$ have equal strength. In general this may not be true. Thus, a general state should be[②]

$$|\psi\rangle = \cos(\theta/2)|\uparrow\rangle + e^{i\phi}\sin(\theta/2)|\downarrow\rangle \tag{6.7}$$

where $0 \leqslant \theta \leqslant \pi$, $0 \leqslant \phi < 2\pi$. Thus a qubit of information spans a sphere, known as the Bloch sphere.

[①] For spin, there are only two "fundamental" states $|\uparrow\rangle$ and $|\downarrow\rangle$ (basis in the quantum sense). The other states, say, $|\rightarrow\rangle$, can be written as a linear superposition of $|\uparrow\rangle$ and $|\downarrow\rangle$. In classical mechanics, we can never imagine that up plus down could be equal to right. This explains ① in Alice's continued adventures.

[②] If we write $|\psi\rangle = \alpha|\uparrow\rangle + \beta|\downarrow\rangle$, we note that normalization of states require $|\alpha|^2 + |\beta|^2 = 1$, and the overall phase is not relevant. Thus we are left with two real parameters.

Spin states in state vectors

Superpositions and linear operators have the math structure of linear algebra. Thus, one can use vectors to represent spin states, and use matrices to represent operators on them. If we choose $|\uparrow\rangle$ and $|\downarrow\rangle$ as the basis, we can write down the states in the z, x, y directions as:

$$|\uparrow\rangle = \begin{pmatrix} 1 \\ 0 \end{pmatrix}, \qquad |\downarrow\rangle = \begin{pmatrix} 0 \\ 1 \end{pmatrix}$$

$$|\rightarrow\rangle = \frac{1}{\sqrt{2}} \begin{pmatrix} 1 \\ 1 \end{pmatrix}, \quad |\leftarrow\rangle = \frac{1}{\sqrt{2}} \begin{pmatrix} 1 \\ -1 \end{pmatrix} \qquad (6.8)$$

$$|\otimes\rangle = \frac{1}{\sqrt{2}} \begin{pmatrix} 1 \\ i \end{pmatrix}, \quad |\odot\rangle = \frac{1}{\sqrt{2}} \begin{pmatrix} 1 \\ -i \end{pmatrix}$$

The analogue for wave functions

Here we are using "matrix mechanics" for spins — find a basis and express things in vectors and matrices. In the part of Quantum Mechanics, we represent states by wave functions. They are equivalent. For the wave function of the position of a particle, we can write $\psi(x)$ as

$$|\psi\rangle = \begin{pmatrix} \vdots \\ \psi(x_1) \\ \psi(x_2) \\ \psi(x_3) \\ \vdots \end{pmatrix}$$

Then up to normalization factors,

$$\langle \psi | \chi \rangle = \int dx\ \psi^*(x)\chi(x)$$

Inner products of spin states

One can define inner products of the state vector. In the matrix sense, if

$$|\psi\rangle = \begin{pmatrix} a \\ b \end{pmatrix}, \qquad |\chi\rangle = \begin{pmatrix} c \\ d \end{pmatrix} \qquad (6.9)$$

define

$$\langle\psi| = \begin{pmatrix} a \\ b \end{pmatrix}^\dagger = (a^*, b^*) \tag{6.10}$$

then

$$\langle\psi|\chi\rangle = (a^*, b^*) \begin{pmatrix} c \\ d \end{pmatrix} = a^*c + b^*d \tag{6.11}$$

Normalization and orthogonality of spin states

The normalization conditions of spin states are

$$\langle\uparrow|\uparrow\rangle = \langle\downarrow|\downarrow\rangle = \langle\rightarrow|\rightarrow\rangle = \langle\leftarrow|\leftarrow\rangle = \langle\otimes|\otimes\rangle = \langle\odot|\odot\rangle = 1 \tag{6.12}$$

And the corresponding pair of states are orthogonal to each other:

$$\langle\uparrow|\downarrow\rangle = \langle\rightarrow|\leftarrow\rangle = \langle\otimes|\odot\rangle = 0 \tag{6.13}$$

All these conditions can be tested from our construction Eq. (6.8).

Operators for spin measurements in z and x, y directions

Given the states which have definite spin (spin eigenstates), one can now construct the spin operators, which measures the spin of the state and gives values ± 1.

One can test that the operators are[1]

$$\sigma_3 = \begin{pmatrix} 1 & 0 \\ 0 & -1 \end{pmatrix}, \quad \sigma_1 = \begin{pmatrix} 0 & 1 \\ 1 & 0 \end{pmatrix}, \quad \sigma_2 = \begin{pmatrix} 0 & -i \\ i & 0 \end{pmatrix} \tag{6.14}$$

These operators correspond to the measurements of spin in the z, x and y directions, respectively. These operators are known as Pauli matrices.

---

[1] The trick for a systematic construction is $\sigma_3 = (+1) \times |\uparrow\rangle\langle\uparrow| + (-1) \times |\downarrow\rangle\langle\downarrow|$, and similarly for σ_1 and σ_2. Here we have defined spin up and down having spin +1 and -1, respectively. In more conventional setup electrons have spins $\pm\hbar/2$. These factors are introduced to compare spin with the familiar concept of angular momentum. Here we will not include these $\hbar/2$ factors.

Operators for spin measurements in a general direction

From Pauli matrices, it is straightforward to build operators for measurements in a general direction $\hat{n}$:

$$\sigma(\hat{n}) \equiv \sum_i \hat{n}_i \sigma_i = \hat{\boldsymbol{n}} \cdot \boldsymbol{\sigma} \qquad (6.15)$$

For example, a measurement along 45° direction of the x-z plane is

$$\frac{\sigma_1 + \sigma_3}{\sqrt{2}} \qquad (6.16)$$

Probability for a measurement to get a particular outcome: z direction

For a spin state $|\psi\rangle$, when measuring the spin of the z-direction, the probability to find the spin up state $|\uparrow\rangle$ is [1]

$$P_\uparrow = \langle \psi | \frac{1 + \sigma_3}{2} | \psi \rangle \qquad (6.17)$$

This is because, the spin state $|\psi\rangle$ can be in general decomposed into

$$|\psi\rangle = \alpha |\uparrow\rangle + \beta |\downarrow\rangle \qquad (6.18)$$

Note that[2]

$$\frac{1 + \sigma_3}{2} |\uparrow\rangle = |\uparrow\rangle \; , \; \frac{1 + \sigma_3}{2} |\downarrow\rangle = 0 \qquad (6.19)$$

Thus

$$\langle \psi | \frac{1 + \sigma_3}{2} | \psi \rangle = |\alpha|^2 \qquad (6.20)$$

which is indeed the probability of finding $|\uparrow\rangle$ according to the measurement postulate.

Probability for a measurement to get a particular outcome: general directions

In general, for the same reason, the probability to find a spin $|\psi\rangle$ along the $\hat{n}$ direction after a corresponding measurement is

$$P_{\hat{n}} = \langle \psi | \frac{1 + \sigma(\hat{n})}{2} | \psi \rangle \qquad (6.21)$$

---

[1] To be more precise, in $1 + \sigma_3$, the 1 is a shorthand of the identity matrix.

[2] Thus $\frac{1 + \sigma_3}{2}$ is the projection operator of the spin-up state.

Pauli matrices acting on spin states

As promised, $|\uparrow\rangle$ and $|\downarrow\rangle$ are eigenstates of σ_3, such that

$$\sigma_3|\uparrow\rangle = |\uparrow\rangle \; , \quad \sigma_3|\downarrow\rangle = -|\downarrow\rangle \qquad (6.22)$$

And it is interesting to note that (as can be tested by direct matrix calculation)

$$\sigma_1|\uparrow\rangle = |\downarrow\rangle \; , \quad \sigma_1|\downarrow\rangle = |\uparrow\rangle \; , \quad \sigma_2|\uparrow\rangle = \mathrm{i}|\downarrow\rangle \; , \quad \sigma_2|\downarrow\rangle = -\mathrm{i}|\uparrow\rangle \qquad (6.23)$$

> **Pauli matrices and rotation**
>
> In addition to the physical meaning of measurements, there is another physical meaning of Pauli matrices: rotation. The σ_1, σ_2, σ_3 Pauli matrices acting on spin states will rotate the spin along the x, y, z axis for 180°, respectively. For example, you can understand (up to a phase) how σ_1, σ_2, σ_3 act on $|\uparrow\rangle$ and $|\downarrow\rangle$ using rotation.
>
> Why Pauli matrices can mean both measurements and rotations? Mathematically, Pauli matrices are both Hermitian (for measurement) and unitary (for rotation as a transformation).

6.2 Multiple Spins and Their Entanglement

6.2.1 Multiple Spins, Entanglements, and Quantum Computing

The last section may not have surprised you since you have already learned some quantum mechanics. However, once we consider two spins, things become surprising again.

Here for simplicity, the two spins are not interacting — there is no force between them. If we were studying classical mechanics, if two subsystems are not interacting with each other, we can just consider them together for free. Once we know both subsystems, we know the whole system. However, in quantum mechanics, things are different due to the principle of superposition.

> **Meaning of states with two arrows**
>
> By writing the state $|ab\rangle$ (where $a,b =\uparrow,\downarrow$), we mean just two states $|a\rangle |b\rangle$ considered together. When taking inner products, we take them separately. For example,
>
> $$\langle cd|ab\rangle = \langle c|a\rangle \langle d|b\rangle$$
>
> And when we act an operator, we will specify the operator acts on which of the two spin. Mathematically, one can generalize the description with the tensor product of matrices, which is beyond the scope of the present course.

States with two spins

In a classical world, if we have two spins, each can be in states $|\uparrow\rangle$ and $|\downarrow\rangle$, when considering them together, we get 4 possible states: $|\uparrow\uparrow\rangle, |\uparrow\downarrow\rangle, |\downarrow\uparrow\rangle, |\downarrow\downarrow\rangle$.

However, in quantum mechanics, superpositions tells that the above 4 states are orthogonal to each other, and thus linear independent. A general state can be written as

$$|\psi\rangle = \alpha_1 |\uparrow\uparrow\rangle + \alpha_2 |\uparrow\downarrow\rangle + \alpha_3 |\downarrow\uparrow\rangle + \alpha_4 |\downarrow\downarrow\rangle \tag{6.24}$$

The state contains two qubits of information.

> **(Optional)More on quantum information**
>
> Here we are talking about "information" in a sloppy way. The discussion can be made precise with a density matrix
>
> $$\rho = \sum_i p_i |\psi_i\rangle \langle \psi_i|$$
>
> where p_i is the classical probability for the system being in state i. Then a whole theory of quantum information can be constructed as a generalization of classical information theory, by considering ρ as a generalization of a list of probabilities. For example, von Neumann entropy $S = -\text{tr}(\rho \ln \rho)$ generalizes the classical Shannon entropy $S = -\sum_i p_i \ln p_i$.

Chapter 6   Entanglement and Quantum Information   175

Entanglement: How much information is contained in two qubits?

We have computed that each qubit is described by $2 \times 2 - 2 = 2$ (two complex parameters deducting normalization and an overall phase) real parameters.

Considering two qubits together, how many real parameters do we need to describe two qubits?

Naively, we may expect: each qubit is described by 2 parameters and thus two qubits would be described by 4. Is it correct?

In Eq. (6.24), there are 4 complex parameters. Deducting normalization and an overall phase, we have $4 \times 2 - 2 = 6$ real parameters.

What are the two unexpected parameters? They describe the entanglement of the two states — more information when considering the two states together than separately. This is a pure quantum effect and thus hard to imagine or explain classically. You can find the math just in the number counting in this subsection, and we will discuss some physics about it in the remainder of this section.

From n spins to quantum computing

How much information is contained in n qubits? We will have 2^n basis vectors and thus a general state is described by $2^{n+1} - 2$ real parameters. This explains observation ② of Alice's continued adventures.

This is exponentially different from classical bits! For n classical bits, the bits can describe *one integer number* with range from 1 to 2^n. But n qubits can describe more than 2^n numbers. Thus, qubits contain much more information in entanglements — the information grows exponentially with the number of qubits.

Further, it is possible to manipulate the qubits with "quantum gates". For example, the Hadamard gate

$$|\uparrow\rangle \to |\rightarrow\rangle \ , \qquad |\downarrow\rangle \to |\leftarrow\rangle \qquad (6.25)$$

and the the "controlled-NOT" gate:

$$|\uparrow\uparrow\rangle \to |\uparrow\uparrow\rangle \ , \quad |\uparrow\downarrow\rangle \to |\uparrow\downarrow\rangle \ , \quad |\downarrow\uparrow\rangle \to |\downarrow\downarrow\rangle \ , \quad |\downarrow\downarrow\rangle \to |\downarrow\uparrow\rangle \qquad (6.26)$$

With quantum information and logical gates to manipulate the information, we get quantum computers!

> **Realistic quantum computing**
>
> Now quantum computers are becoming real. But building a quantum computer challenges the science and technology today. For a quantum computer to work, one needs:
>
> (1) Construct the qubits and keep them coherent (maintain the information from quantum entanglement).
>
> (2) Construct quantum gates to operate on qubits.
>
> (3) Design mechanism to correct possible errors from realistic operation.
>
> (4) Design quantum algorithms. See Quantum Algorithm Zoo for currently known quantum algorithms.

Measurements and corresponding operators

For multiple qubits, here we limit our attention to measurements which operates separately on each individual qubit. And for simplicity we consider 2 qubits only.

The independent measurements can be represented by two sets of Pauli matrices, one acting on particle 1 (not affecting particle 2) and the other acting on particle 2 (not affecting particle 1). Let us use σ_i to denote the Pauli matrices on particle 1, and use τ_i to denote the Pauli matrices on particle 2. For example,

$$\sigma_1 |\uparrow\downarrow\rangle = |\downarrow\downarrow\rangle \,, \qquad \tau_1 |\downarrow\downarrow\rangle = |\downarrow\uparrow\rangle \tag{6.27}$$

As σ_i and τ_i act on different spin states, the order of operation σ_i and τ_i commutes. For example, $\sigma_1 \tau_1 |\psi\rangle = \tau_1 \sigma_1 |\psi\rangle$.

As the two measurements are independent, the probability to find particle 1 in direction $\hat{n}$ and particle 2 in direction $\hat{m}$ (by measuring corresponding directions) in a state $|\psi\rangle$ is

$$\langle\psi| \frac{1+\sigma(\hat{n})}{2} \frac{1+\tau(\hat{m})}{2} |\psi\rangle \tag{6.28}$$

6.2.2 The Einstein-Podolsky-Rosen (EPR) Paradox

Einstein, though a pioneer of quantum mechanics, has a life-long concern about quantum mechanics. In his philosophy, "God does not play dice." He did not believe that a theory having fundamental randomness would be the complete description of nature. Along this line of thought, in 1935, Einstein, Podolsky and Rosen (EPR) posed the strongest challenge to quantum mechanics.

The EPR state

Let us first prepare a state (known as the "singlet state")

$$|s\rangle = \frac{1}{\sqrt{2}} (|\uparrow\downarrow\rangle - |\downarrow\uparrow\rangle) \tag{6.29}$$

and take the two particles far apart each other. This is possible. For example, a spin-less particle at rest can decay into two particles with opposite spins, and these two particles can fly as far apart as one wants, before they interact with anything else.

The correlated measurements

Suppose Alice and Bob are light years away from each other. The decay event happens in their middle. Later, Alice receives particle 1 and Bob receives particle 2. They measure the spin of the particles along the z-direction immediately after receiving the particles. What will they see?

We already have the theory to answer this question technically:

$$\begin{aligned} P_{\uparrow\uparrow} &= \langle s| \frac{1+\sigma_3}{2} \frac{1+\tau_3}{2} |s\rangle = 0 \,, & P_{\downarrow\downarrow} &= \langle s| \frac{1-\sigma_3}{2} \frac{1-\tau_3}{2} |s\rangle = 0 \\ P_{\uparrow\downarrow} &= \langle s| \frac{1+\sigma_3}{2} \frac{1-\tau_3}{2} |s\rangle = \frac{1}{2} \,, & P_{\downarrow\uparrow} &= \langle s| \frac{1-\sigma_3}{2} \frac{1+\tau_3}{2} |s\rangle = \frac{1}{2} \end{aligned} \tag{6.30}$$

In words, both Alice and Bob will find that they have equal chance to find the state having spin up or down. It seems random. But if they meet and compare notes, they will find that their results are totally *correlated* — if one gets up, the other must get down.

It is interesting to note that, the above correlation does not depend on the direction that Alice and Bob measure the state. As long as they are measuring the same direction (any $\hat{n}$), if one gets up, the other must get down. This correlation (without any interactions) is the physical indication of entanglements.[①]

[①] This explains observation ③ of Alice's continued adventures.

Does quantum mechanics contradict relativity?

At first sight, in EPR, information is transmitted faster than light. Thus it contradicts causality in special relativity. This is the EPR paradox.

However, there is actually nothing wrong. Before Alice and Bob meet, they consider their own results *random*. Thus Alice cannot encode any information (classical information) in what she choose to measure and let Bob know immediately.

Is EPR trivial?

There are many apparent superluminal correlations, which are trivial. For example,

① When a child is born, his/her father immediately becomes a father no matter how far away he is.

② I pack an apple and an orange into the two boxes. Alice and Bob randomly pick one of them. Then they separate in space. Their measurement about what's in the box will have complete correlation.

Clearly, these types of correlations cannot send information either. However, is EPR similar to these types of trivial correlations?

EPR is not any of them.

① is more trivial. There is correlation, but the correlation is not observable (but rather just a definition) before the father gets any information about the birth of the child.

② is not what happened in EPR, either. In the case of ②, Alice and Bob have mutual information,[①] which is similar to the case of EPR. However, the mutual information is classical information, instead of quantum entanglement.

If Alice choose to measure the spin along the x axis, but Bob still measure long z axis, then their results will not have any correlation (or any further correlations no matter they choose to measure whatever in addition). Whether their results have correlation depends on what they choose to measure.

You may not feel satisfied with this explanation — by measuring different axes, you may argue that one of Alice and Bob destroyed the correlations (recall the collapse of wave function). So this argument is trying to hide (destroy) the difference between EPR and apple-orange instead of trying to compare them honestly.

[①] The mutual information in the apple-orange case is $I(X;Y) \equiv \sum_{x,y} p(x,y) \log_2 \frac{p(x,y)}{p(x)p(y)} = \log_2 2$. This is just one bit of information in information theory (where the base of log is usually chosen to be 2).

Also, what if we generalize ②? The generalization along this line is known as "local hidden variable theories". They are theories with hidden variables which are not (yet) observable.

Is it possible to design a better local hidden variable theory (where the randomness of quantum mechanics is due to some variables not-yet-observable), where the "random" outcome of EPR (and similarly other quantum measurements) is in fact determined when the two spins were together? And after the states fly apart, each state follow classical logic separately?

We will show in the next subsection that EPR is indeed intrinsically different from apple-orange in ②, and further, no local hidden variable theories can exist.

Classical hidden variables

There are lots of randomness in the classical world due to conditions out of control. For example, when a dice is played, the outcome is (classically) in principle determined by initial condition of the dice, perturbations in the air, mechanical property of the desk, etc. But it's too hard to understand these variables to predict the behavior of the dice. Quantum randomness is fundamentally different, as we will see later.

6.2.3 The Bell's Inequality and Its Violation

Defining a local hidden variable theory in general

A local hidden variable theory need to satisfy the follows:

(1) *Hidden variables*. The apparent quantum behavior follows from some classical logic (where some parts of the system may be hidden so the rest appear random).

(2) *Local*. Space-like measurements do not affect each other.

A piece of classical logic

Let us prepare many copies of identical EPR pairs. We classify each EPR pair using three classification criteria of states A, B and C.

One consequence of the classical logic is:

$$N(A, \neg B) + N(B, \neg C) \geq N(A, \neg C) \tag{6.31}$$

where $N(...)$ denotes the number of states satisfying condition ..., and $\neg$ denotes "not". This relation is illustrated in Fig. 6.3, where A, B, C are denoted by circles and the colored contours denote corresponding terms in the above equation.

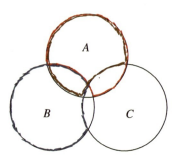

Fig. 6.3

If you consider classical objects and assign, e.g. A: apple (instead of orange); B: red (instead of yellow); C: sweet (instead of sour). Then Eq. (6.31) will always be satisfied. What about EPR pairs with quantum entanglement?

Testing the classical logic upon the singlet state

Now let us give meanings of A, B and C, applying on a singlet state, as measurements by Alice:

A: Measure particle 1 along the z-axis and get positive spin.

B: Measure particle 1 along 45° in x-z plane and get positive spin.

C: Measure particle 1 along the x-axis and get positive spin.

Interestingly, $\neg A$, $\neg B$ and $\neg C$ can be related to measurements by Bob of particle 2 because of the complete anti-correlation:

$\neg A$: Measure particle 2 along the z-axis and get positive spin.

$\neg B$: Measure particle 2 along 45° in x-z plane and get positive spin.

$\neg C$: Measure particle 2 along the x-axis and get positive spin.

Eq. (6.31) with the above conditions is known as the Bell inequality (John Stewart Bell 1964).

Now it is straightforward to construct probabilities of the events in Eq. (6.31). The probabilities become number of events after a large number of measurements, on many copies of EPR pairs.

$$P(A, \neg B) = \langle s | \frac{1+\sigma_3}{2} \frac{1+\frac{\tau_1+\tau_3}{\sqrt{2}}}{2} | s \rangle = \frac{1}{4}\left(1 - \frac{1}{\sqrt{2}}\right) \simeq 0.073 \qquad (6.32)$$

$$P(B, \neg C) = \langle s| \frac{1 + \frac{\sigma_1 + \sigma_3}{\sqrt{2}}}{2} \frac{1 + \tau_1}{2} |s\rangle = \frac{1}{4}\left(1 - \frac{1}{\sqrt{2}}\right) \simeq 0.073 \qquad (6.33)$$

$$P(A, \neg C) = \langle s| \frac{1 + \sigma_3}{2} \frac{1 + \tau_1}{2} |s\rangle = \frac{1}{4} = 0.25 \qquad (6.34)$$

Clearly, ① the classical logic Eq. (6.31) is violated in this state. Thus the quantum information stored in the entanglement cannot be mimicked by any local hidden variables.

> **Why EPR violates Bell inequality?**
>
> By violating Eq. (6.31), we are not saying that EPR pairs are not logical. Rather, we cannot assume that each particle in the EPR pair already has features A, B and C before the measurement. This is totally different from the apple-orange example.

6.3 Epilogue: Summary and What's Next

Further reading

(1) The main reference of this part is Susskind and Friedman, *Quantum Mechanics: The Theoretical Minimum*.

(2) For more information about spin: Sakuri, *Modern Quantum Mechanics*, Chapter 1.

(3) For similar content and other Bell-like experiments: Baumann, *Lecture Notes on Modern Physics*, Chapter 2.

(4) If you'd like to dive deeper in quantum information, read the lecture notes by Preskill.

---

① This explains observation ④ of Alice's continued adventures.

What happens next?

Quantum information and quantum computing are fast developing frontiers in research. It is definitely interesting to study them further in quantum mechanics and related courses.

Exercises

E6.1 Spin eigenstates

Find the eigenstates and eigenvalues of $\sigma(\hat{n})$.

E6.2 Computational details for the Bell inequality

Compute the probabilities $P(A, \neg B)$, $P(B, \neg C)$ and $P(A, \neg C)$.

Chapter 7
From the Action to the Laws of Nature

A new way of thinking about physics

We are used to Newton's way to think about physics — for a particle, by knowing the initial position and velocity, we calculate the trajectory of the particle as time evolves. This way of thinking is generalized to gravity, E&M fields, and so on. The equation that governs the time evolution is known as the equation of motion.

The equation of motion

The equation to determine the particle's trajectory is known as the equation of motion, which is usually a second order differential equation (or a set of such equations) with time. For example, the motion of a particle in a potential $V(q)$ can be described by an equation of motion $\ddot{q} + dV/dq = 0$. To solve a realistic problem, the equation of motion is also packed with initial conditions: at an initial time t_0, the values of $q(t_0)$ and $\dot{q}(t_0)$. To solve an equation of motion with such initial conditions is called a Cauchy problem.

These are all good. But if we really want to ask fundamental questions — can these equations of motion be the ultimate codes of nature — there are still a few unpleasant aspects:

① Equations of motion are not invariant objects in special relativity. Because an equation of motion is time evolution. And in relativity time and space appears with almost equal rights. For different observers, equations of motion are covariant (transforms consistently with Lorentz transformation), but not invariant. Is the ultimate code of nature really so subjective? In the movie "Matrix", the world is a program. Has the programer chosen a preferred frame in this movie to write down equations of motion and code the world?

② Where do conservation laws come from? We are aware of energy, momentum, angular momentum, charge conservation laws. They can be proven given a particular framework of the theory. For example, given Newtonian mechanics we can prove energy and momentum conservations by smart tricks. But is there a universal way to figure out the origin of conservation laws for a general system?

③ Motion with constraints. For example, think about a pendulum (if it's too simple, think about double pendulum). In an ideal theory, constrained motion means that we are reducing possibilities and thus the situation should be simpler. But in the Newtonian way, the more constraints, the more analysis of forces and the more annoying math. Is there a way to make constrained systems simpler to compute?

④ The quantum world is so different from its classical counter part.

⑤ The equations of motion are too "cold blooded" to describe how nature "would like to" behave. Can you use one sentence to conclude the nature of Nature, which applies to every known piece of fundamental physics? For example, "Nature prefers to …", saying, Nature prefers to satisfy its equation of motion does not seem pleasant enough.

A short sentence by Maupertuis (1698—1759) addresses all the above issues:

"Nature is thrifty in all its actions."

We will see how this simple sentence work.

7.1 Fermat's Principle of Light

Before addressing the action principle, let us first use the Fermat's principle to study the propagation of light. This is not the real action principle yet, but the physical concept and math method are very similar. At the same time it is more intuitive. So let's first do this warm up exercise here.

> **Pierre de Fermat**
>
> Pierre de Fermat (1601/1607—1665) is a French lawyer. Nobody remembers what case he has defended. But we do remember many things about him, including his conjecture that there is no positive integer solution for $x^n + y^n = z^n$ for $n > 2$. After 300 years, this was finally proven in 1994.

Fermat's principle

Consider a class of light propagation problems: Given two fixed points P and Q, how does light propagate between those points?

Some examples are given in Fig. 7.1, including propagation in free space, reflection and refraction.

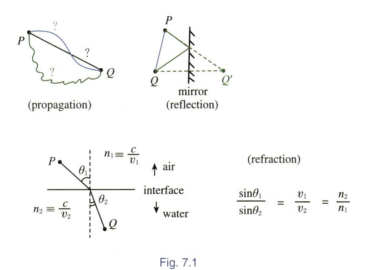

Fig. 7.1

Fermat proposed a general solution for this type of problems: Light travels between two given points along the path of extremal time.

How travel time depends on path? Functionals

You are familiar to the problems of extremal problem in calculus: A function $f(x)$ takes extremal value when $df/dx = 0$. Here our situation is similar, but with a different mathematical object.[①] We are talking about the space (in math sense) of "path". How to parameterize a path in math? A path can be described as a function. For example, a curve $y = y(x)$.

Suppose we know the speed of light (c in vacuum, c/n in media with reflection index n). Given a path, we know the light propagation time T. Here T depends on the whole function y (i.e. not only the value of $y(x)$ at a particular x). We then say that T is a *functional* of y, denoted as $T[y]$ with square brackets.

In short, a functional is a "function of function". A comparison of functions and functionals are sketched in Fig. 7.2.

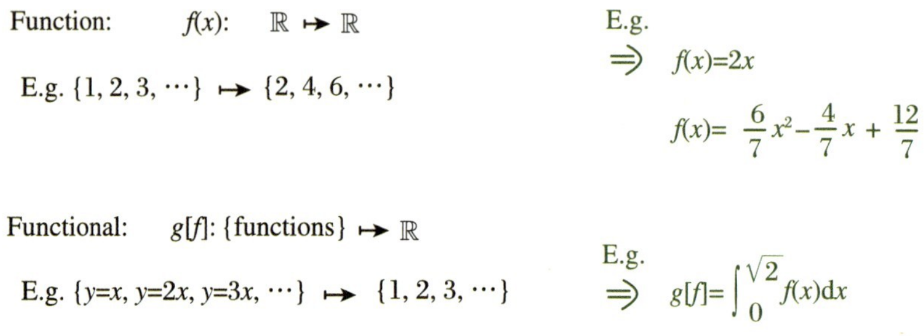

Function: $f(x)$: $\mathbb{R} \mapsto \mathbb{R}$ E.g.
$$\Rightarrow f(x) = 2x$$
E.g. $\{1, 2, 3, \cdots\} \mapsto \{2, 4, 6, \cdots\}$
$$f(x) = \frac{6}{7}x^2 - \frac{4}{7}x + \frac{12}{7}$$

Functional: $g[f]$: $\{\text{functions}\} \mapsto \mathbb{R}$ E.g.
$$\Rightarrow g[f] = \int_0^{\sqrt{2}} f(x) dx$$
E.g. $\{y=x, y=2x, y=3x, \cdots\} \mapsto \{1, 2, 3, \cdots\}$

Fig. 7.2

Functional programming

In programming language there is a paradigm known as functional programming. The very basic requirement is that you can pass a function as an argument of another function. It is the same functional (function of function) there.

Which path has extremal time? Functional variation

To find extremal time between (x_P, y_P) and (x_Q, y_Q), we first draw a (general) path $y(x)$ satisfying $y(x_P) = y_P$ and $y(x_Q) = y_Q$ (Fig. 7.3). Then we find the condition that a extremal path must satisfy with the following procedure:

We vary the path by $y(x) \to y(x) + \delta y(x)$, where δy is an *arbitrary* infinitesimal function (i.e. we ignore $(\delta y)^2$) satisfying $\delta y(x_P) = 0$ and $\delta y(x_Q) = 0$. These "boundary con-

---

[①] Of course not all paths can be parametrized by $y = y(x)$, for example x =constant. But here let's restrict our attention to paths which can be described by $y = y(x)$, which provides us enough background to proceed to the action principle.

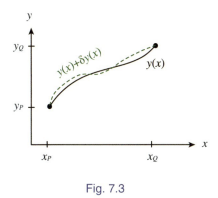

Fig. 7.3

ditions" are needed because we have fixed these two boundary points by the definition of the problem.

We can now write the time of the new path as $T[y+\delta y]$, and thus the functional variation

$$\delta T \equiv T[y+\delta y] - T[y] \tag{7.1}$$

A path with extremal time must have $\delta T = 0$ for all the possible variations $\delta y(x)$. The above seems dry and empty. Let's consider an example.

Why $\delta T=0$ means extremal?

The reason is similar to why an extremal point of a function $y(x)$ has $dy/dx = 0$ in calculus. Let's say if the extremal is a local minima. Let's consider the change of $y(x)$ within $(x, x+dx)$, keeping $y(x)$ outside this region fixed (this is one allowed form of δy). Then the change will be $\delta T = (\ldots)\delta y dx$, if $\delta T \neq 0$ within $(x, x+dx)$. If $(\ldots)$ is negative, the change δy makes T smaller and thus T is not minimal. If $(\ldots)$ is positive, the change $-\delta y$ makes T smaller and thus T is not minimal. Thus, for minimal T, δT vanishes within $(x, x+dx)$. Since $(x, x+dx)$ is a general interval, δT should vanish everywhere. The same argument applies for local extrema.

Light propagation in the vacuum

We study light freely propagating in the vacuum between (x_P, y_P) and (x_Q, y_Q). Suppose we only know the Fermat principle, and that the speed of light is c. For simplicity, we suppress the z direction and only study x, y spatial dimensions. The propagation time

T between this two points can be written as

$$T = \int_{t_P}^{t_Q} dt = \frac{1}{c} \int_{x=x_P, y=y_P}^{x=x_Q, y=y_Q} \sqrt{dx^2 + dy^2} = \frac{1}{c} \int_{x_P}^{x_Q} \sqrt{1 + \left(\frac{dy}{dx}\right)^2} \, dx \qquad (7.2)$$

To find the condition that an extremal path $y(x)$ must satisfy, we vary $y(x) \to y(x) + \delta y(x)$ and insert it into the above equation. As $d\delta y/dx$ is small, we can consider it as a small parameter and do Taylor expansion:

$$\sqrt{1 + \left(\frac{d[y(x) + \delta y(x)]}{dx}\right)^2} = \sqrt{1 + (y')^2} + \frac{y'}{\sqrt{1 + (y')^2}} \frac{d\delta y}{dx} + \mathcal{O}[(\delta y)^2] \qquad (7.3)$$

where $y' \equiv dy(x)/dx$. Thus

$$\begin{aligned}
\delta T[y] &= \frac{1}{c} \int_{x_P}^{x_Q} \frac{y'}{\sqrt{1 + (y')^2}} \frac{d\delta y}{dx} \, dx \\
&= \frac{1}{c} \int_{x_P}^{x_Q} \left\{ \frac{d}{dx}\left(\frac{y'}{\sqrt{1 + (y')^2}} \delta y\right) - \delta y \frac{d}{dx}\left(\frac{y'}{\sqrt{1 + (y')^2}}\right) \right\} dx \\
&= -\frac{1}{c} \int_{x_P}^{x_Q} \delta y \frac{d}{dx}\left(\frac{y'}{\sqrt{1 + (y')^2}}\right) dx \qquad (7.4)
\end{aligned}$$

where we have dropped the total derivative term because $\delta y = 0$ on the boundary. $\delta T = 0$ requires that the above equation vanishes for all δy. Thus we require

$$\frac{d}{dx}\left(\frac{y'}{\sqrt{1 + (y')^2}}\right) = 0 \quad \to \quad y' = \text{const} \qquad (7.5)$$

We have thus proven that light travels with straight lines from the Fermat's principle.

With more efforts, one can also derive the light trajectory for the reflection and refraction cases using Fermat's principle and variations. But we shall not do it here.

Here we are doing something trivial using complicated methods. But the techniques (math) developed here will be reused in the next section for the action principle.

7.2 Principle of Extremal Action

One may live a life in two ways: (1) given the current state, this is what I'd like to do now. And I will evolve with time and see what future will become; and (2) given the current state, I have a clear dream of my future. And I will find out the master plan where my dream can come true.

Interestingly, physics can also be understood in these two ways (and they are equivalent): (1) corresponds to the Cauchy problem — solving the equation of motion with initial position and velocity; and (2) corresponds to the action principle, which we are now to introduce.

The action principle

A theory is defined by an action S. The equation of motion of the theory corresponds to the extremal action $\delta S = 0$.

Let us first see how this works in Newtonian mechanics. Later we generalize it to include all known fundamental theories.

Action principle in Newtonian mechanics

Newtonian mechanics can be formulated as the following question:

Given a potential $V(q)$, where $q(t)$ is the position of the particle, what is the trajectory between a starting point (q_1, t_1) and a final point (q_2, t_2) (Fig. 7.4)?

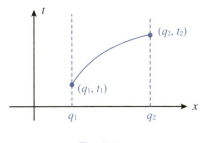

Fig. 7.4

First let us define an action

$$S[q] = \int_{t_1}^{t_2} dt \left[\frac{1}{2} m \dot{q}^2 - V(q) \right] \qquad (7.6)$$

The action has a form of integrating $K - V$ over time, where K is the kinetic energy and V is the potential energy. This is quite general. Do not ask about the physical meaning at the moment. We just define such a functional (functional of paths $q(t)$) and see what it leads to. We will come back to its physical meaning (and also why there is an extremal action principle at all) in Section 7.4.

> **A remark on integral convention**
>
> Physicists sometimes write $\int dt f(t)$. This is identical to $\int f(t) dt$. It is just a matter of notation. But it reflects that physicists tend to think an integral as a limit of summation: $\lim_{\Delta t \to 0} \sum_i \Delta t f(t_i) = \int dt f(t)$. Here Δt and $f(t_i)$ are just multiplied together and thus the order is not important.

To derive the equation of motion, similarly to Section 7.1, we vary $q(t) \to q(t) + \delta q(t)$, which satisfies $\delta q(t_1) = \delta q(t_2) = 0$, and see how the action changes:

$$\delta S = \int_{t_1}^{t_2} dt \left[m\dot{q}\delta\dot{q} - \frac{dV}{dq}\delta q \right] = \int_{t_1}^{t_2} dt \left[\frac{d}{dt}(m\dot{q}\delta q) - m\ddot{q}\delta q - \frac{dV}{dq}\delta q \right] \tag{7.7}$$

Again the first term is a boundary term that vanishes. The last two terms holds for all $\delta q(t)$ and thus we have re-derived Newtonian equation of motion for a particle:

$$m\ddot{q} + \frac{dV}{dq} = 0 \tag{7.8}$$

> **Motion with constraints**
>
> By now, you should be able to solve the motion with constraints, for example, double pendulum in a smarter way than Newtonian analysis of forces. As an exercise, you may use two angles to parameterize K and V of the system and write $S = \int dt(K - V)$. Variation principle gives you much simpler result than what you would have calaulated using forces.

A General action and the Euler-Lagrange equation

In general, consider a "Lagrangian" $L = L(q_i, \dot{q}_i, t)$, where $i = 1, 2, \ldots, N$. Here q_i denotes the position of the i-th particle. (In fact, in the spirit of general coordinate, the index i can also collectively denote many possible things: different spatial dimensions, different particles, values of fields at a point, and so on. We will not dive into those details.)

The action is defined as

$$S = \int L(q_i, \dot{q}_i, t) \mathrm{d}t \tag{7.9}$$

Here we have not written down the limits of the integral, having in mind that the boundary terms will be dropped. Clearly, this definition includes the Newtonian particle example that we have studied, and is much more general.

> **Math of the Lagrangian**
>
> By writing $L = L(q_i, \dot{q}_i, t)$, it is in the sense of multi-variable calculus: one can calculate partial derivatives on the three variables "independently" assuming the rest two variables do not change. For example, $\partial_t L \equiv \partial L / \partial t$ assumes that q_i and $\dot{q}_i$ do not change when calculating the partial derivative. This is different from
>
> $$\mathrm{d}L/\mathrm{d}t = \sum_i (\partial L/\partial q_i)(\mathrm{d}q_i/\mathrm{d}t) + \sum_i (\partial L/\partial \dot{q}_i)(\mathrm{d}\dot{q}_i/\mathrm{d}t) + \partial L/\partial t$$

The equation of motion of such a general system, known as the Euler-Lagrange equation, can be derived by the variation

$$\begin{aligned}\delta S &= \sum_i \int \left(\frac{\partial L}{\partial \dot{q}_i} \delta \dot{q}_i + \frac{\partial L}{\partial q_i} \delta q_i \right) \mathrm{d}t \\ &= \sum_i \int \left[\frac{\mathrm{d}}{\mathrm{d}t}\left(\frac{\partial L}{\partial \dot{q}_i} \delta q_i \right) - \frac{\mathrm{d}}{\mathrm{d}t}\left(\frac{\partial L}{\partial \dot{q}_i} \right) \delta q_i + \frac{\partial L}{\partial q_i} \delta q_i \right] \mathrm{d}t \end{aligned} \tag{7.10}$$

Thus the Euler-Lagrangian equation (holds for every i) can be read off as

$$\frac{\mathrm{d}}{\mathrm{d}t}\left(\frac{\partial L}{\partial \dot{q}_i} \right) - \frac{\partial L}{\partial q_i} = 0 \tag{7.11}$$

> **Dropping the boundary temrs**
>
> From now on we will assume that the boundary terms $\int \partial_t(...) \mathrm{d}t$ can be dropped. The argument is that this term does not modify the EoM. A careful treatment of the boundary terms is beyond the scope of this lecture; but dropping these boundary terms doesn't hurt for all our current purposes.

All known fundamental physics in one line

All known fundamental physics can be written into an action

$$S \sim \int d^4x \sqrt{-g} \left\{ \frac{1}{16\pi G} R - \frac{1}{4} F^2 + i\bar{\psi} D\psi + |Dh|^2 - V(h) + h\bar{\psi}\psi \right\} \tag{7.12}$$

This action covers the laws of gravity (the metric g), E&M and their friends (gauge field in F), electrons and their friends (spinor fields represented by ψ) and origin of mass (the Higgs field h). These fields, put together with certain gauge symmetries, is known as the particle physics Standard Model and describes all known fundamental physics. We shall not explain it here (a course of particle physics or quantum field theory needed). Take it as a piece of art at the moment.

7.3 Symmetry and Conservation Laws

In the chapter of Special Relativity, at relativistic momentum and energy, we have argued why we need conservation laws: (1) physically, allow us to ignore the details happened in the middle; (2) mathematically, reduce 2nd order ODEs into 1st order or 0th order; (3) provide observables for modern physics.

But the question left is, why there exists conservation laws after all? To ask the question in general, we had better to use a universal framework of physical theories to find the root of conservation laws. Thus the general action (7.9) is a good starting point.

Instead of directly asking why anything is conserved in (7.9), let's first think about the beauty of a theory— symmetry of a theory.

A symmetry of a theory

A symmetry of a theory is a transformation under which the theory does not change under.

In terms of Eq. (7.9), the above definition of symmetry can be put more explicitly as: Consider an infinitesimal transformation

$$q_i \to \tilde{q}_i = q_i + \epsilon \delta q_i \tag{7.13}$$

where ϵ is an infinitesimal constant. If the action does not change under

$$\delta S = \int L(\tilde{q}_i, \dot{\tilde{q}}_i, t) \mathrm{d}t - \int L(q_i, \dot{q}_i, t) \mathrm{d}t = 0 \tag{7.14}$$

then we consider the transformation (7.13) as a symmetry.

> **Remarks about symmetry**
>
> The action is a means to derive equations of motion. Thus to examine the change of the action under Eq. (7.13), any equation of motion must not be used (to avoid circular arguments). To be explicit, when talking about $\delta S = 0$, we may mean one of the two things: (1) a symmetry transformation without using equation of motion; or (2) any transformation (may not be a symmetry) as a variation principle, and use the equation of motion. One should be clear about the difference.
>
> Again we assume to drop boundary terms $\int \partial_t(\cdots) \mathrm{d}t$ when considering the change of the action. This is to say, if the Lagrangian $L(\tilde{q}_i, \dot{\tilde{q}}_i, t) \neq L(q_i, \dot{q}_i, t)$, but rather $L(\tilde{q}_i, \dot{\tilde{q}}_i, t) = L(q_i, \dot{q}_i, t) - \epsilon \mathrm{d}g/\mathrm{d}t$ for arbitrary g, the transformation is still considered a symmetry.
>
> As ϵ is infinitesimal, we will ignore the $\mathcal{O}(\epsilon^2)$ terms.
>
> The type of symmetry Eq. (7.13) that we study here is a continuous and global symmetry. There are other types of symmetries, namely discrete symmetries and gauge symmetries, which will not be useful to derive nontrivial conservation laws.

The definitions of Eq. (7.13) and Eq. (7.14) are very abstract. We thus consider a few examples.

Examples of transformations

(1) Time translation, to test if the prediction of a theory is time dependent. The transformation thus relates two situations: doing an experiment now and doing an experiment a bit later (within the same theory) and observing if there is any difference. The time translation can thus be written as

$$q_i(t) \to \tilde{q}_i(t) = q_i(t + \epsilon \delta t) = q_i(t) + \epsilon \dot{q}_i(t) \delta t \tag{7.15}$$

We thus extract that for time translation

$$\delta q_i = \dot{q}_i(t) \delta t \tag{7.16}$$

We will later study the consequence of time translation in great detail in this section.

(2) Space translation[①], to test if doing an experiment in one place is identical to doing an experiment in a slightly different location. The transformation is thus $q_i(t) \to q_i(t) + \epsilon \delta q_i$, where δq_i is a constant shift. The space translation can be related to momentum conservation.

(3) Lorentz transformation. The boost between (t_B, x_B) and (t_A, x_A) for infinitesimal β (or rotation). When β is small, $\gamma \approx 1$. Then approximately $t_B \simeq t_A + \epsilon v x_A/c^2$, $x_B \simeq x_A + \epsilon v t_A$. We thus have $q_i(t) \to q_i(t + \epsilon v q_i/c^2) + \epsilon v t$. This boost part can be related to (not very interestingly) the initial center of energy. The rotation part of Lorentz transformation can be defined similarly and is related to angular momentum conservation.

(4) Change of phase. If q_i are complex, it makes sense to examine the transformation $q_i \to e^{i\epsilon\alpha} q_i$. This can be related to charge conservation.

We have studied many transformations. Let's move on to explore the requirement for a transformation to be a symmetry. We only take time translation as an example.

When is time translation a symmetry?

Intuitively, if the theory as specified by the Lagrangian L does not explicitly depend on t (i.e. $L(q_i, \dot{q}_i, t) = L(q_i, \dot{q}_i)$, without explicit t dependence), the theory is time translation invariant. Here we will test that in this case, the time translation is indeed a symmetry.

Under time translation, considering that ϵ is a constant, we have

$$\delta q_i = \dot{q}_i \delta t, \qquad \delta \dot{q}_i = \ddot{q}_i \delta t \tag{7.17}$$

$$\delta L = L(q_i + \epsilon \dot{q}_i \delta t, \dot{q}_i + \epsilon \ddot{q}_i \delta t) - L(q_i, \dot{q}_i) = \left[\frac{\partial L}{\partial q_i} \dot{q}_i + \frac{\partial L}{\partial \dot{q}_i} \ddot{q}_i\right] \epsilon \delta t = \epsilon \frac{\mathrm{d}(L\delta t)}{\mathrm{d}t} \tag{7.18}$$

When ϵ is a constant (as we already defined), this is indeed a symmetry. To see that, note that a total derivative is a boundary term (which we neglect) in the action. Thus (7.14) holds.

We have inserted examples above to make things not too dry. But now let us back to very general discussions following Eq. (7.13) and Eq. (7.14). They are not limited to time translation but in general for any symmetry.

[①] For space translation, is δq_i the same for different i? We will need to examine what i actually means then: If i stands for diffferent directions in space, then δq_i can take different values. If i stands for different particles in the same direction, then δq_i must take the same value for different i.

Conservation laws from symmetries

Bear with me a mathematical trick: We have defined ϵ as a constant. Now, let's vary it: $\epsilon = \epsilon(t)$. A symmetry transformation keeps the action invariant when $\epsilon = $ const. Now how should it change when $\epsilon = \epsilon(t)$? The change of action should now take the form

$$\delta S = \int (P\epsilon + Q\dot{\epsilon})dt = \int Q\dot{\epsilon}dt = -\int \dot{Q}\epsilon dt + \text{(neglected boundary terms)} \quad (7.19)$$

Here the P term vanishes because a symmetry requires this term to vanish when taking $\epsilon = $ const (or P may contain a total derivative part, which can be absorbed to Q after an integration by parts). There are no terms such as $\ddot{\epsilon}$ because $L = L(q, \dot{q}, t)$ and there is no $\ddot{q}$ to generate $\ddot{\epsilon}$. At the last step integration by parts is used. ①

Now we are ready to interpret the physical meaning of Eq. (7.19). If we allow to use equations of motion, that is to say, $\delta S = 0$ for all possible $\epsilon(t)$ (not because of a symmetry, but because of the action principle). Thus we have $\dot{Q} = 0$. In other words, Q is a conserved quantity when equations of motion are used.

This correspondence between a symmetry and a conserved quantity is known as the Noether theorem.

In the above, we have shown the existence of a conserved quantity without showing actually what it is. This is not our style (unless you are a mathematician). What is the form of a conserved quantity given L and δq?

What is the conserved quantity in general?

We have shown that, a symmetry leaves the action invariant and thus change the Lagrangian by at most a total derivative: $\delta L = -\epsilon dg/dt$ for some quantity g. When $\epsilon = \epsilon(t)$, we cannot drop this term. Also, the change of allowing $\epsilon = \epsilon(t)$ adds a term in the following step:

$$\delta L = \sum_i \left[\frac{\partial L}{\partial q_i} \epsilon \delta q_i + \frac{\partial L}{\partial \dot{q}_i} \partial_t(\epsilon \delta q_i) \right] \supset \sum_i \frac{\partial L}{\partial \dot{q}_i} \delta q_i \dot{\epsilon} \quad (7.20)$$

They are all the new terms that $\epsilon = \epsilon(t)$ brings and thus

$$\delta S = \int \left(g + \sum_i \frac{\partial L}{\partial \dot{q}_i} \delta q_i \right) \dot{\epsilon} dt \quad (7.21)$$

---

① If you really want to be careful about the boundary terms of the action, here you can choose $\epsilon(t)$ being a function that vanishes on the initial and final boundaries and thus we can indeed get rid of boundary terms.

where we have performed an integration by parts to the g-term. Compare with Eq. (7.19), we have the conserved quantity

$$Q = g + \sum_i \frac{\partial L}{\partial \dot{q}_i} \delta q_i \qquad (7.22)$$

When the equation of motion is used, $\dot{Q} = 0$ (conservation law).

> **When can EoM be used**
>
> We did not allow to use the equations of motion (EoM) when testing if a transformation is a symmetry. But now, for requiring a conserved quantity to conserve, the equation of motion can indeed be used. For example, energy conservation indeed needs Newton's 2nd law (or relativistic generalizations) to apply. If a particle accelerates freely as it likes, it does not conserve energy.

What is Eq. (7.22)? For different symmetries, Q stands for different conserved quantities. Let's see one example: time translation in Newtonian mechanics.

Example: energy conservation as a result of time translation symmetry

Under time translation, the change of Lagrangian is Eq. (7.18). Thus, $g = -L\delta t$. The conserved quantity is the "Hamiltonian"

$$H \equiv \frac{Q}{\delta t} = \sum_i \frac{\partial L}{\partial \dot{q}_i} \dot{q}_i - L \qquad (7.23)$$

What's this? Consider more explicitly

$$L = \sum_i \frac{1}{2}\dot{q}_i^2 - V(q_1, \ldots, q_n) \qquad (7.24)$$

The conserved quantity is thus

$$H = \sum_i \frac{1}{2}\dot{q}_i^2 + V \qquad (7.25)$$

This conserved Hamiltonian is indeed the energy of the system.

> **When is conservation broken?**

The Noether theorem does not only tell what is the conserved quantity, but also tell explicitly the quantity is conserved under which situation. In the past we state that in an isolated system energy is conserved. Now we know that

(1) The condition can be relaxed: even the system is not isolated, as long as the time translation symmetry is not broken by the environment, energy is still conserved. (For example, in the case with a fixed gravitational potential, consider the source of gravity, such as the earth to be outside the system.)

(2) If the system is time-dependent, even if the system is isolated, energy may not be conserved. An example is our expanding universe.

7.4 The Hidden Quantum Reality

The nature seems "strange"

The action principle offers a new way of thinking about how physics works. Take for example the motion of a particle moving from A to B in a force field (Fig. 7.5):

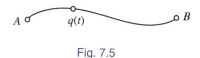

Fig. 7.5

(1) Newton's view: The initial position and velocity are given. At each moment, the particle "feels" the force and "adjusts" its velocity according to the force. The adjusted velocity "tells" the particle how to move further.

(2) The action principle: The starting and end points are fixed, the particle needs to "find" its way between these two points based on an extremal action.

Do you feel the action way "stranger"? It's straightforward for a particle to "adjust" its velocity in a force field (Newton's view). However, a particle cannot calculate (seriously in the theory of computation, since it may not even carry a bit of information or complicated enough for being even one logical gate). So how a particle can actually "follow" the action principle?

Moreover, as we argued the action principle is a more fundamental way to describe all laws of nature. Why nature behaves fundamentally in such a "strange" way?

The nature is natural but quantum

No. The nature is not strange. The nature is natural, but just not classical.

Let's return to the question of the particle motion. Instead of "calculating" the extremal action, what the particle actually does is that, it "tries" **all possible paths** and "take" the extremal one.①

The above words are actually nothing but the path integral formulation of quantum mechanics. In quantum mechanics, the probability P for things to happen (here: particle to propagate from A to B) is decomposed as

$$P = |\mathcal{A}|^2 \qquad (7.26)$$

where the complex number $\mathcal{A}$ is known as the probability amplitude. The probability amplitude can be calculated by a weighted average over all paths

$$\mathcal{A} \propto \sum_{\text{all paths}} e^{iS/\hbar} \qquad (7.27)$$

Here by "all paths" we mean all possible lines connecting A and B, which not necessarily satisfy the equation of motion.

What if $S \sim \hbar$?

What about $S \sim \hbar = 1.05 \times 10^{-34} \cdot \text{m}^2 \cdot \text{kg} \cdot \text{s}^{-1}$? Let us jump into the brave new world of quantum mechanics to answer this question in the next part! But in fact, we will use different but equivalent formulations of quantum mechanics (mainly wave mechanics and a bit of matrix mechanics) instead of the path integral. This is because the path integral, though conceptually simple, is harder in many computations. It will be included in a full quantum mechanics course.

---

① You may think about a similar question: Why in a circuit, the electric current "knows" which way to go to minimize power? No, the current doesn't know. Rather, the electric field build up all possible paths until reach a stationary situation. More obvious examples include why flood "knows" where to flow, and so on. Definitely they are classical. But a particle can rely on its quantum nature to achieve a similar feature.

The action principle explained

In the quantum world, the physical meaning of the action is clear: $e^{iS/\hbar}$ is the phase factor as a weight of the path in the summation of all paths.

You may be confused here: in classical mechanics, the particle only select one trajectory. In quantum mechanics (path integral formulation), the particle moves along all possible paths together. How to reconcile the difference? How does the classical trajectory emerge among the quantum trajectories?

Classical mechanics emerges from quantum mechanics when $S \gg \hbar$. In this limit, even a very small change in S (due to choosing a nearby different trajectory) results in a large change in the unit $\hbar$, and thus $e^{iS/\hbar}$ is a fast oscillating function, which cancels contributions of different paths almost for all paths.

There is only one exception: close to the stationary action $\delta S = 0$. Near $\delta S = 0$ different paths do not cancel. Thus the classical trajectory is the only trajectory left over when the classical limit $S \gg \hbar$ is taken.

7.5 Epilogue: Summary and What's Next

Further reading

(1) Similar contents in other textbooks: You can find an extensive discussion of the content here in Part II of *Einstein Gravity in a Nutshell* by Zee. You may also read the first section and the final appendix of *Lecture Notes on Modern Physics* by Baumann.

(2) If you want even more references, I recommend to watch *Theoretical Minimum* (Video Lectures) (Lectures 3 and 4 of *Classical Mechanics*) by Susskind; or to read Chapter 2 of *Classical Mechanics* by Goldstein, Poole Jr and Safko.

What happens next in a university physics program?

(1) Classical mechanics. The action principle is closely related to the Lagrangian formulation of classical mechanics. There is also a Hamiltonian formulation. You will see how mechanics are written in these ways. You will learn how Lagrangian mechanics is

helpful in solving constrained motion problems.

(2) Quantum mechanics. The path integral mentioned here will be part of quantum mechanics (or sometimes advanced quantum mechanics). The (quantized) Hamiltonian is also a centural part of quantum mechanics.

(3) Quantum field theory. A model in quantum field theory starts with an action. You will fully see there that the action principle is indeed considered as the first principle.

Exercises

E7.1   Refraction from the Fermat's principle

Derive the law of refraction from the Fermat's principle.

E7.2   Extremal but non–minimal paths

Find examples that light travel along extremal, but not minimal paths.

E7.3   A cosmological scalar field

In cosmology, a homogeneous and isotropic scalar field has action

$$S = \int dt \, a^3(t) \left(\frac{1}{2}\dot{\phi}^2 - V(\phi) \right)$$

where $a(t)$ is a function of time, and $V(\phi)$ is a function of ϕ. Calculate explicitly the Euler-Lagrange equation of ϕ (i.e. relation between $\ddot{\phi}$, $\dot{\phi}$ and ϕ) in two ways: The Euler-Lagrange equation and the variation principle, respectively.

E7.4   From Euler-Lagrange to Newton

Start from the Euler-Lagrange equation, use the Lagrangian of a particle in Newtonian mechanics, to derive the Newtonian second law for particle motion in a potential.

E7.5 A relativistic free particle

The action for a freely moving (i.e. not moving in a force field or interaction with other particles) relativistic particle is

$$S = \alpha \int d\tau = \alpha \int \sqrt{1 - \frac{\dot{q}^2}{c^2}}\, dt \tag{7.28}$$

where α is a constant.

(1) Determine the value of α by taking the Newtonian limit $\dot{q} \ll c$.
(2) Show that the system has time translation symmetry.
(3) Derive the relativistic energy as the conserved quantity of time translation.

Chapter 8
From Particles to Strings

8.1 Elementary Particles

We have understood for long that atoms or nuclei are not the most fundamental particles in the nature. What are the more fundamental constructs of nature?

Point particles

So far, we have not detected any size or shape of elementary particles. In other words, they behave as point particles. Having said that, keep in mind that we are talking about point particles in the quantum sense: the particle can be in superposition of position eigenstates, and thus appear extended, thanks to the particle-wave duality. However, if we measure the position of the particle, the particle appears as a point.

Elementary particles

Along the way of subatomic structures, the nuclei are made of protons and neutrons. The protons and neutrons are made of quarks. To the best of our current knowledge, there are some types of point particles that appears most elementary, and we have not observed their substructures. What are they? And what are their natures?

Fundamental matter (fermions)

Leptons: electron (e), μ and τ.
Neutrinos: ν_e, ν_μ, ν_τ.
Quarks: u, c, t, d, s, b.

Fundamental forces (bosons)

E&M force: photon.
Strong force: gluon.
Weak force: W, Z.
Gravity: graviton.

These particles also have their anti-particles. In addition, there is a "Higgs boson", responsible for the origin-of-mass of the fundamental particles, sometimes also classified as a kind of force.

> **What's the origin of mass?**
>
> Mass is one of the most important nature of elementary particles. However, the mass of a particle may not be fundamental. Where does mass come from?
>
> Elementary leptons and quarks: their mass comes from their interaction with the Higgs. By interacting with the Higgs background, these particles can no longer move at the speed of light, and thus become massive. This is known as the Higgs mechanism.
>
> Nuclei: nuclei mass dominates the mass of atomic matter. The mass of the nuclei is dominated by the binding energy between the quarks through $m = E/c^2$, instead of the mass of the quarks themselves.
>
> The Higgs mass seems to have its own origin. We don't know either the origin of neutrino mass and dark matter mass, either.

The theory to describe these particles and their interactions is the standard model of particle physics.

However, these known matter components only consists of about 5% of the energy content of our universe. The rest are about 25% dark matter and 70% dark energy. We do not know what they are in the particle physics sense.

Also, gravity is not well-described in the particle physics standard model. It is believed that the standard model is an effective theory instead of a final theory. At even shorter length (beyond current experiments), new physics should arise. For example, more types of particles, or even extending the point particles to one-dimensional objects known as strings.

Who ordered that?

Almost all the matter that we see in the world are made of electrons, and the u and d type quarks (making up atoms).

Thus, it was surprising when more elementary particles were discovered. For example, when μ was discovered from cosmic rays, behaves like a heavier brother of electron, Rabi quipped "who ordered that?"

The matter particles other than e, u or d are not seen in our everyday lives because:

(1) Some are dead. The μ, τ leptons and c, t, s, b quarks are unstable and will decay in microscopic time scales.

(2) The rest are shy. The neutrinos ν_e, ν_μ, ν_τ interact with us too weakly, and thus we don't feel them. For example, every second, there are 100,000,000,000,000 neutrinos (mainly from the sun) passing through your body. However, among them, only 10^{-8} neutrino interact with you (in the probability sense), i.e. only of order 10 neutrinos interact with you throughout your life. You certainly can not feel that 10 atoms in your body moved by these neutrinos. Dark matter are probably also shy particles, which may be the reason why we haven't seen them directly.

For a similar reason, we are more familiar with E&M and gravity, than the mysterious weak and strong forces. We do not see the weak force in everyday life for a similar reason as we don't see neutrinos: the weak force is too weak. And we do not see the strong force because it is too strong: Free quarks cannot appear in our everyday energy scale. The strong force confines the quarks in pairs or triples, and thus their long range communications are not seen. It's like if a couple always appear together, you will not see them exchange messages by emails.

The ultimate goal of elementary particle physics is to put everything into the same framework for study. This effort is known as unification. Unification has been one of the strongest driving forces for physics. Have we unified everything into the same work of elementary particle physics?

Quest for unification

One of the most amazing power of physics is unification. Unification brings apparently unrelated things together; reduce the number of laws of nature; and even extend the arenas of physics. For example,

(1) Newton: the terrestrial and celestial bodies appear so different and thus Aristotle suggested that they should be described by different laws. However, Newtonian's Law of Universal Gravitation unified the two classes of bodies.

(2) Maxwell: the electric and magnetic phenomena appear so different. However, Maxwell unified them into electromagnetic fields by Maxwell equations and light emerges there. Later, the 4-dimensional spacetime formalism further manifest the unification of electric and magnetic fields.

(3) Einstein: space and time appear so different, acceleration and gravity appear so different, fluctuation and dissipation appear so different...

(4) Grand unification: the electromagnetic, weak and strong forces are conjected to be unified at an energy scale over 10^{10} higher than our current experimental capabilities. Due to the experimental limitations, we don't know so far if grand unification is indeed the origin of the electromagnetic, weak and strong forces.

(5) Supersymmetry: bosons and fermions look so different. But even them may be put into the same framework. The transformation relating bosons and fermions is known as supersymmetry. So far we still don't know if supersymmetry indeed exists in nature, or it is just a beautiful mathematical structure that nature did not use.

So far so good. The particle physics standard model is a successful framework to describe all known fundamental particles. Although as of writing, we are still not sure about the naturalness of Higgs mass, origin of neutrino mass, the nature of dark matter, and the asymmetry between matter and anti-matter, there are many proposals for solving these problems. People tend to believe that the solution of these problems would be within the same framework that we understand our familiar fundamental particles.

However, does that mean that everything has been put into the same unified framework?

No. We have not mentioned gravity yet. Can gravity be put into the same framework of quantum mechanics? Modern gravity is described by general relativity, which is one of the most elegant theory we have ever seen. However, it appears isolated from the kingdom of elementary particle physics. How to unify gravity with the rest part of laws of physics?

8.2 Quantum Gravity

Quantum gravity is widely believed to be the most important, and probably also the most difficult question in theoretical physics. The problem of quantum gravity remains an open question. Nevertheless, tremendous progress has been made. In this section, we will see why quantum gravity is so difficult, and outline some possible ways out.

8.2.1 Do We Need Quantum Gravity, and Where to Find It?

Before to proceed to quantum gravity, let us first step back and ask, whether gravity can remain classical.

Can gravity remain classical?

Can gravity remain classical, while other kinds of matter are quantized? For example, what if we let gravity couple to the expectation value of quantum matter density $\langle \rho \rangle$ and pressure $\langle p \rangle$, and so on? It is in general believed not an option. Arguments for the necessity of quantum gravity include:

(1) Gravity suffers some same problems with why light should be quantized. For example, the ultraviolet catastrophe for the statistical mechanics of a gas of gravitational waves.

(2) In general, it is difficult to couple a quantum system to a classical system in a fundamental way. For example, if gravity couples to expectation values in quantum mechanics, then do all gravitational forces indicate measurements, which collapses quantum wave functions? (i) If all gravitational forces are considered measurements of the system: ruled out by the Colella-Overhauser-Werner experiment, in which the superposition of matter affected by a gravitational field is observed. (ii) If some gravitational forces are not considered as measurements: in principle, their reaction can be measured, from which we

can get more information than allowed to violate the uncertainty principle (though this is still too difficult to observe in experiments). Also, it's very hard, if possible at all, to preserve the linearlity of quantum mechanics while coupling it to a classical system.

(3) Black hole entropy. If gravitational waves were classical, whose energy is not limited by the quantizaton condition $E \geqslant h\nu$, by sending low energy gravitons into a black hole, we would have decreasing entropy and violate the second law of thermodynamics.

(4) The region where GR is incomplete largely coincide with that of quantum gravity. This is not a coincidence. In general, at high energy scales, general relativity is ill-behaved and physical quantities diverge. For example, black hole singularities and cosmic big bang singularities are believed to be related to quantum gravity.

(5) Action principle. Gravity can be elegantly described by the action $\int d^4x \sqrt{-g} R$. Quantum mechanics should be the underlying reason why the action principle works.

So classical gravity is not an option. Thus we have to dive into quantum gravity in our search of a more fundamental theory (if not the most fundamental) of nature. Let's see how difficult it is, and find possible ways out. Where to find quantum gravity?

Planck units from dimensional analysis

The natural scale of quantum gravity was first proposed by Planck in 1899. At that time Planck was on his way to launch the new era of quantum mechanics and already noted the importance of a new constant of nature $h \approx 6.63 \times 10^{-34}$ m² · kg · s⁻¹, now known as the Planck constant. Planck not only noted the role of h in a new interpretation of black body radiation, but also noted that hidden in h there is a natural way to measure things — not natural in human's convenience, but in the way that nature indicates, without referring to any scale specially indicated by human.

Before Planck, units are man-made. You can define a person's foot to be the length unit — but why not another person? Can we bootstrap units without artificial impacts?

Planck noted that a set of units hide in the constants of nature, h, the speed of light $c \approx 3.00 \times 10^8$ m/s, and the Newtonian gravitational constant $G \approx 6.67 \times 10^{-11}$ N · m² · kg⁻². It's as simple as primary school math:

$$l_\mathrm{P} \equiv \sqrt{\frac{Gh}{c^3}} \simeq 4.05 \times 10^{-35} \text{ m}, \qquad t_\mathrm{P} \equiv \sqrt{\frac{Gh}{c^5}} \simeq 1.35 \times 10^{-43} \text{ s},$$
$$m_\mathrm{P} \equiv \sqrt{\frac{hc}{G}} \simeq 5.46 \times 10^{-8} \text{ kg}, \qquad E_\mathrm{P} \equiv \sqrt{\frac{hc^5}{G}} \simeq 4.90 \times 10^9 \text{ J} \qquad (8.1)$$

where l_P is known as the Planck length, t_P is known as the Planck time, m_P is known as the Planck mass and E_p is known as the Planck energy.

> **Reduced Planck units**
>
> Sometimes the reduced Planck constant $\hbar \equiv h/(2\pi)$ is used instead of h in the definition. The resulting units are known as the reduced Planck length, time, mass and energy, respectively.

For the first time, units emerges from laws of nature, without any scale specified by human (although there may be some $\mathcal{O}(1)$ parameters still relying on human's conventions in physical formulae). What is the physical meaning of the Planck units?

> **How hard to get there?**
>
> Noted that in weight, m_P is about 0.05 mg. Isn't this amount of matter what a chemist deals with every day?
>
> If you are only talking about the number, right. However, here we are talking about *one fundamental particle* with mass 0.05 mg, not many particles together. To compare, an electron weights $\sim 10^{-24}$ mg, the heaviest fundamental particle known now is the top quark, whose mass is $\sim 3 \times 10^{-19}$ in the unit of mg.

Physical meaning of Planck units

Considering a quantum elementary particle with mass m. In quantum mechanics, the particle has an intrinsic wave length — the Compton wave length $\lambda = h/(mc)$, which is the minimal uncertainty of the particle's position (we can make the position more certain by using higher energies to probe the particle, but at the expense of creating anti-particles and the breakdown of the single-particle picture).

Can we infinitely decrease this minimal uncertainty of the particle's position by increasing it's mass m? We note that there is a fundamental limit preventing us to reducing the uncertainty of the particle's position to zero — once the particle reaches Planck mass m_P, the compton wavelength reaches (of order) the Schwarzschild radius. Beyond this point, further increasing m, you get a larger black hole, whose horizon prevents you to locate the particle.

In short, if $m \sim m_P$, gravitational effects of this particle become strong. Quantum effects and gravitational effects of this particle must be considered at the same time —

this is the scale of quantum gravity.

Thus, now we know the scale of quantum gravity. Can we simply put gravity into the framework of quantum mechanics, similar to what we have done with electromagnetism (electrodynamics + quantum = quantum electrodynamics, a well-established theory already in the 1950s)?

8.2.2 Theoretical Challenges

Interaction between matter and quantum fluctuations

The vacuum is full of quantum fluctuations. How do quantum fluctuations contribute to propagation of particle, or particle interactions? Consider the following example:①

Here a particle is propagating (Fig. 8.1). It interacts with quantum fluctuations (the dashed line, either quantum fluctuation of the same particle type, or another type of particles). The quantum fluctuation car-

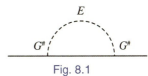

Fig. 8.1

ries energy E. Note that quantum fluctuation with all possible energies and momenta can occur. And thus we have to integrate over all energy and momentum. Since we will compare the case with/without quantum fluctuations, we expect that the total quantum fluctuations can be expressed in a dimensionless number Δ:②

$$\Delta = \int dE \, d^3p \text{ (quantum fluctuations with energy } E \text{ and momentum } p) \qquad (8.2)$$

Unfortunately, this integral may not converge in the $E \to \infty$ and $|p| \to \infty$ limits (known as ultraviolet divergence, or UV divergence for short). This has puzzled physicists multiple times and has led to breakthroughs such as quantum electrodynamics (renormalize model parameters to absorb divergences), divergence as a guidance of flow of theories in theory space (renormalization group), naturalness as a guidance for physical discoveries, and the key feature of quantum gravity. Quantum theories (to be more precise, quantum field theories) falls into the following 4 kinds depending on the convergence behavior of the above integral:

---

① Note that quantum corrections can occur in all manners, and what we plotted is only one simple example. There can be quantum correction to particle interactions, or multiple quantum corrections happened together. Can you figure out what they look like?

② Alert readers may note that here we have integrated E and p independently without noting the relativistic energy-momentum relation $E^2 = p^2c^2 + m^2c^4$. There are two formulations to calculate these quantum corrections: on-shell (satisfying the energy-momentum relation) or off-shell formulations. Here we use the off-shell formulation, where quantum fluctuations do not have to satisfy the energy-momentum relation.

① Finite: no divergence appears. If so, the theory do not intrinsically need new physics at high energies. But Unfortunately, ultraviolet finite theories are very rare, and usually require symmetries not yet observed in nature, such as supersymmetries and conformal symmetries.

② Renormalizable: all the divergences can be absorbed into redefinition of model parameters. The appearance of divergence indicates that the theory has to be ultraviolet-completed at high energy. But we can safely use the theory as a self-contained theory without knowing when the theory would break down. Renormalizable theories further falls into classes of natural/unnatural, which we will not further discuss here.

③ Non-renormalizable: to absorb more and more divergences, one has to extend the model and introduce more and more parameters. The theory is still effective (as an effective field theory) at low energies, but breaks down at a predictable scale.

Gravity falls into category ③. At low energies, we can trust classical gravity or even do a little of perturbative quantum gravity by considering small quantum corrections. However, when quantum corrections become important, the theory breaks down. What is the scale where our gravity theory breaks down? We can estimate that from dimensional analysis. Each gravitational coupling brings a positive power of G. Thus, if we cut off the upper limit of the integral (8.2), $\int^\infty dE \to \int^\Lambda dE$ and similar for $|p|$, we have $\Delta \sim (G\Lambda^2/(\hbar c^5))^n$. And the more complicated the diagram is, we get the more powers of n. Clearly, all these corrections break down at Planck energy $\Lambda = E_P$.

Thus, we can still study gravity as a low energy effective theory and even study small quantum corrections. But if we would like to understand a full theory of quantum gravity, allowing large quantum contributions, we have to understand gravity at the Planck scale. This is a difficult job. We have mentioned some difficulties. Now let us summarize them, mention a few more, and they show why quantum gravity is difficult.

Crazy things at the Planck energy scale

(1) Non-renormalizability: infinitely many divergences appear and cannot be absorbed into redefinition of model parameters.

(2) The Compton wavelength of a particle is comparable of the Schwarzschild radius.

(3) Spacetime background fluctuate so strongly that we cannot treat space and time as smooth background parameters. Even the causal structure may not be preserved.

(4) Time is an important parameter in quantum mechanics, and in general relativity, it is a coordinate parameter which can be reparametrized.

(5) Black hole entropy indicates that on the black hole horizon, on average, each

Planck area stores of order one bit of information. But we do not understand how the information is encoded for realistic black holes.

(6) Definition of observables. In particle physics, we can let particles collide — in this setup we can define free particles when the particles are far away. Thus the interaction rate between the particles is a well-defined observable (cross-section and S-matrix). But with gravity, the presense of horizons (say, cosmological horizon) may not allow particles to be far enough to be free.

Is difficulty disgusting?

Difficulty is usually disgusting. However, the theoretical difficulty of quantum gravity may be an exception. This is because, due to experimental difficulties, quantum gravity already lacks experimental guidance. If the theory of quantum gravity were easy, we may be left too many possibile theories. On the contrary, the difficulty of quantum gravity does not allow so many possible quantum gravity theories. Some physicists even believe that there is a unique self-consistency theory of quantum gravity, which can be found out by self-consistency without much experimental hints.

Is self-consistent quantum gravity theoretically unique? We don't know yet. If it is indeed unique, we should thank to the difficulty of quantum gravity.

A zoo of proposals of quantum gravity

There are many proposals to solve the problem of quantum gravity. We are not sure which is the right theory. The proposals of quantum gravity include:

(1) *String theory* asserts that the world is made of one dimensional strings (and other extended objects) instead of point particles. Applying rules of quantum mechanics to strings, gravity emerges. This is the leading theory of quantum gravity, at least in terms of the size of the community. We will discuss this in the next section.

(2) *Loop quantum gravity (LQG)* seeks for the fundamental degrees of freedom in general relativity to quantize. In the early era of LQG, the Wilson loop of the connection was considered the fundamental excitations. Later, more development indicates that the fundamental degrees of freedom of gravity may be the holonomy of certain gravitational connection and the flux of triad.

(3) *Asymptotic safety* examines flows of theories as energy scale varies (known as renormalization group flow), and conjectures that theories involving gravity flows to a fixed region at high energy, where the divergent behavior of gravity becomes milder.

There are many other approaches, such as causal set theory, dynamical triangulation, and so on.

8.2.3 Experimental Challenges

Which proposal of quantum gravity to follow? This should be a question to be answered by experiments. Unfortunately, experiments involving quantum gravity are exceptionally hard. In this subsection, we will show the difficulty, possibilities and progress towards experimentally probing quantum gravity.

Where to find quantum gravity?

To talk about quantum gravity, we need to first identify where to look for quantum gravitational effects. There are three possibilities:

① Experiments to directly probe the Planck scales. This is ideal, but too difficult to be done at any foreseeable future.

② Experiments to reach as large energy as we can, and use precision measurements to search for quantum gravity effects of fundamental particles (usually suppressed by $(E/E_p)^2$ or more, for experiments with experiments with energy scale E).

③ Thanks to the advancement in quantum technology, now superposition of bigger and bigger systems are possible. For example, even superpositions of bacteria are being discussed. More matter means more gravity. Can we probe quantum effects of gravity in this way? Note that since in this approach, the energy scale E is usually low, we are not likely to be exploring the new physics associated with quantum gravity by $(E/E_p)^2$ corrections.

In the remainder of this subsection, we will outline experiments according to the above classification ①, ② and ③ of quantum gravity effects.

Strong gravity: how large collider to build?

Without considering technical limitations, in theory, the most ideal way to probe Planck scale physics would be to build a particle collider, which can accelerate particle to the Planck energy E_p. To reach E_p, how large collider do we need?

To simplify the problem, we limit our attention to a toy linear collider, with a uniform electric field E accelerating an electron.

Can we make the electric field E infinitely strong? Unfortunately, not even theoretically. This is because, electron-positron pairs emerge if the electron field is strong enough to separate vacuum fluctuations. This mechanism is known as Schwinger pair production. The work to separate electron-positron pairs from vacuum fluctuations is $W \sim Ee\lambda_c \sim m_e c^2$, where m_e is the electron mass and $\lambda_c = h/(m_e c)$ is the Compton wavelength of the electron. Thus, the maximal electron field $E = m_e^2 c^3/(eh) \simeq 2.1 \times 10^{17}$ V/m, and to reach Planck energy, the length L of the linear collider should satisfy

$$M_P c^2 \sim EeL, \qquad L \sim \lambda_c \frac{M_P}{m_e} \sim 10^{11} \text{ m} \sim 100 \ R_{\text{sun}} \sim 1 \text{ AU} \qquad (8.3)$$

In words, to reach the Planck scale, a linear collider has to be more than 100 times longer than the radius of the sun.

Weak gravity: hints for quantum gravity?

Cosmology of the very early universe may be an arena in searching for quantum gravity. Since it is generally believed that the universe has reached a state with much higher energy density and temperature than that of any man-made experiments. So far we have not detected a signal. But active progress is made in many directions, for example,

(1) Primordial gravitational waves. Searching for relic primordial gravitational waves with cosmic wavelength can be think of as searching for gravitons (which was amplified by the expansion of the universe, and then leave signature relics on the cosmic microwave background). As of writing, many experiments are making fast progress in this direction.

(2) Cosmological collider with high spin. Through density correlations in our universe, one can search for relics of early universe interactions involving spin-two particles or higher. If there is such a discovery, it would indicate not only quantum gravity, but new physics arising from quantum gravity.

(3) A package of predictions from quantum gravity models of the very early universe. For example, as a string cosmological model, brane inflation may be verified by a package of predictions such as preferred appearance of the two-point density correlation function and cosmic string productions.

Even weaker: can we detect a graviton?

Can we build a detector which, when one graviton flies by, there is a high probability to detect it? How to convert this question to semi-quantitative estimations?

We can calculate the mean free path L of the graviton. If the graviton has large probability to be detected by the detector, the length of the detector should be at least of order this mean free path.

How to calculate the mean free path? $L = 1/(\sigma n)$, where n is the number density of particles in the detector, and σ is the cross section, i.e., how large the graviton appears in area (in an effectively classical ball collision sense). We may estimate $\sigma \sim l_P^2$, for example from graviton interacting with relativistic matter, where no other scales comes in (except some dimensionless parameters such as $\alpha \sim 1/137$, which is not very significant here). ①

How to choose the particle number density? Even if we make the particles as dense as a neutron star, we still get the mean free path of the graviton to be 10^{25} m, i.e., 1% of the diameter of the whole observable universe (which spans 900 light years). This detector is clearly too long to make. Not to mention that so much neutron star matter will fall into a black hole when put together.

Alternatively, we may use black holes to absorb gravitons. This is much more efficient due to the infinitely deep gravitational potential. However, how do we know the black hole has absorbed a graviton? We need extremely precise measurement of the change of black hole mass, which again is an extremely difficult gravitational measurement.

Thus, detecting a single graviton with large probability is extremely difficult. What about processes producing a huge number of gravitons and we try to detect one per century? Some estimates② show that one would need to put a Jupiter-mass detector in the orbit of a compact object such as a neutron star for this purpose.

Superposition of bigger objects

Recently, there is a growing interest of quantum gravity through superposition of large objects, thanks to the rapid development in quantum information. The question here is superposition (or entanglement) v.s. gravity. Some possible interplays between superposition and gravity include:

(1) Superposition affected by classical gravity. This is not surprising by modern standard. And experiments has already be made since 1975 by Colella, Overhauser and Werner.

(2) Entanglement caused by gravity.

(3) The gravitational field created by superposition of matter.

---

① The estimation $\sigma \sim l_P^2$ is hand-waving here, and can actually follow from a careful calculation. See, for example, arxiv:gr-qc/0601043. A naive non-relativistic calculation gives $\sigma \sim l_P^4/\lambda_c^2$, where λ_c is the Compton wavelength of the matter interacting with the graviton, which would have resulted in much greater difficulty in detecting a graviton.

② https://arxiv.org/pdf/gr-qc/0601043.pdf

The latter two proposals are not realized by experiments at the moment, but they or similar ideas may be realized in the foreseeable future.

8.3 Is the World Made of Strings?

Now let us return to the theoretical difficulty of quantum gravity. Many difficulties point to the wild ultraviolet behavior of quantum gravity. How to make gravity milder?

Extended objects and milder gravity

How to make quantum gravity less divergent in the ultraviolet? There are many proposals. Among them, the most intuitive idea is perhaps to generalize point particles into extended objects. In this way, energy density gets smoothed, and thus spacetime geometry gets smoothed at small scales.

To make this idea explicit, we illustrate particle interactions in Fig. 8.2. Ultraviolet divergence happens when the intermediate particles have extremely short wavelengths. Now that the particles are made of extended objects, the divergence is removed.

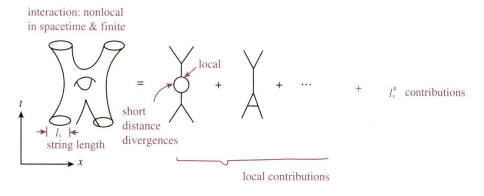

Fig. 8.2

How was string theory born?

The history of string theory is a bit like quantum mechanics. Everything started in 1968 when Veneziano proposed a formula to fit strong interaction — recall that in 1900 Planck proposed a formula to fit black body radiation. In 1970, Nambu, Nielsen and Susskind understood the physical meaning behind Veneziano's formula: it arises if the fundamental particles are actually little strings.

Later, string theory did not go far to explain strong interaction. One reason is that unwanted spin-two particles automatically arise in string theory. The spin-two particles are unwanted for strong interaction, but that is exactly what quantum gravity needs! This starts the journey to explore whether string theory would be the theory of everything.

Dimensionality of the extended objects

Now that we accept the idea of extending fundamental particles into higher dimensional objects, how many dimensions should these objects have?

Once the fundamental objects are made extended, we must consider the quantum mechanics of their internal degrees of freedom as well. The quantum fluctuations of the vibration modes of the object can be characterized similar to Eq. (8.2), but replacing the dimension of the integral with $\int dE d^n p$, where n is the internal spatial dimension of the particle (instead of spacetime dimension, technically, $n+1$ is the dimension of the object's "world volume").

(1) If $n=0$, we return to the limit of a point particle, where we do not worry about its internal vibration at all, but this is nothing new and we do not solve the problem of quantum gravity.

(2) If $n=1$, the particles are extended to strings. This is the balance where in the spacetime theory, gravity emerges (surprisingly) and is finite; while the internal dynamics of the string is also mild enough.

(3) If $n>1$, from the geometry of the shape of the extended object, its own "gravity" emerges (recall that gravity is geometry in general relativity) and become non-trivial. Before worrying about the spacetime gravity, we need to first worry about the gravity of the extended object itself.

Thus, it's natural to proceed with the possibility of $n=1$, in which the fundamental objects are strings with one spatial dimension. Indeed, in the non-perturbative dynamics of string theory, higher or lower dimensional objects, such as D-branes emerge. But this is beyond the scope of the introduction here.

Cool ideas needed by string theory

(1) Supersymmetry: Are bosons and fermions symmetric? They appear so different. But in 1960s-70s, some theories were proposed in which bosons and fermions can transform to each other keeping the theory invariant. In string theory, supersymmetry is needed for the stability of strings' ground state.

(2) Extra dimensions: can we have more than 3 space dimensions? In 1919, Kaluza proposed that there may exist extra dimensions which are small and thus we do not see them. And Maxwell's electromagnetism emerges when reducing high-dimensional gravity to lower dimensions. The theoretical consistency of the theory indicates that there has to be 9 space dimensions + 1 time dimension in string theory.

We haven't observed supersymmetry or extra dimensions yet. How do we interpret the need of these ideas? Does that mean string theory is so great that these beautiful ideas emerge, or we have to involve these complexities to save string theory by hand? We don't know so far.

Cool results arising from string theory

(1) Quantum gravity: we have already mentioned the difficulty of quantizing gravity. Amazingly in string theory quantized gravity arises for free. Whether or not is string theory the theory of the real world, now it is widely considered as at least a self-consistent model of quantum gravity. Putting gravity into the framework of a quantum theory can also be seen as a step towards unification.

(2) Grand unification: apart from gravity, the rest 3 fundamental interactions of nature are electromagnetic, weak and strong. Can they be unified? This leads to grand unification theories, which can be embedded into string theory.

(3) Dualities: appearently there are multiple versions of string theory. For example, whether you choose to allow strings to form closed loops only, or the two ends of a string can be open. Surprisingly, there are dualities to relate those theories — these theories are equivalent and it is believed that there is a unique version of string theory in the non-perturbative sense.

(4) Uniqueness: as hinted by dualities, starting from the postulate that the world is made of strings, we may arrive at a unique theory.

(5) Landscape: although string theory is believed to be unique, some estimates show that when reduced to low energy, string theory may have $10^{100} \sim 10^{500}$ solutions. This is like that governed by the same universal gravity, asteroids can have all kind of shapes. These

googols of solutions allow rich phenomena of low energy laws of nature as we experience.

The uniqueness and landscape of string theory might have gone too far. They are not only the scientific features of string theory. Rather, if we take string theory seriously, these uniqueness and landscape features may reshape our understanding of science. We end up this part by more comments on these two natures of string theory.

Uniqueness: a new kind of science?

Fundamentally, string theory may be unique. Science is grounded in observations and experiments. But uniqueness of string theory, allows researchers to, for the first time, proceed so far theoretically beyond the reach of experiments (recall the difficulties of observationally test quantum gravity). Is it great progress, or we have been marching on a wrong way? We don't know. Eventually, we need to find experimental predictions to tell.

Landscape: a new kind of science?

Effectively, string theory may have googols of solutions. This challenges our fundamental methods of scientific interpretation.

For example, why the fine structure constant is $\alpha \approx 1/137$? Why dark energy takes 70% of the energy of our universe? Traditionally minded, we would find a theory to interpret it. But with googols of solutions in the most fundamental theory, we may have to give up traditional types of explanations because in the other solutions these numbers may well be different.

More exotically, the existence of intelligent beings (say, us) may play a role — in these googols of solutions, only in the solutions allowing intelligent beings, scientific questions can be asked. So some fundamental constants of nature may be explained in a similar way as why the earth temperature allows liquid water — otherwise we are not there to ask this question. This reasoning is known as the anthropic principle, which was proposed in the 1970s and made popular by the modern understanding of string theory.

Do we live in a string landscape? Do some constants of nature have to be interpreted by the anthropic principle? We don't know.

8.4 Epilogue: Summary and What's Next

Further reading

There is a series of "DeMYSTiFieD" books, among which the quantum field theory and string theory volumes tries to introduce relevant topics in the most accessible way. There is also a book *A First Course in String Theory* and courses based on that. String theory is also a popular topic of popular science. There are many great popular science books, for example, *The Elgant Universe: Superstrings, Hidden Dimensions, and Quest for the Ultimate Theory*.

What's next?

We don't know.

Chapter 9
Entropy and Information

In thermodynamics, you were told by a law that "entropy" increases, in order to forbid the dream of a perpetual motion machine. To be more explicit, for an isolated system consists of n subsystems, for each subsystem i define

$$\Delta S_i = \frac{\Delta Q_i}{T} \tag{9.1}$$

where ΔQ_i is the heat flowing into the subsystem if the process is reversible. Then for a general process, the entropy of the system (Clausius 1855) satisfies the second law of thermodynamics:[①]

$$\Delta S = \sum_{i=1}^{n} \Delta S_n \geqslant 0 \tag{9.3}$$

The equal sign is taken for reversible process; and for non-reversible processes $S > 0$. Also, S is a quantity describing the state of the system.

Instead of repeating the standard discussion in thermodynamics here (I assume that you know them. Otherwise see further readings), let us think about a few questions left unclear:

[①] And recall that the first law is

$$dE = dQ - pdV \tag{9.2}$$

The mysterious entropy in thermodynamics

① S describes the state of the system. Which aspect of the system does S describe?[1]

② Fundamental laws of nature has no arrow of time in it. How can anything increase, which generates an arrow of time?

③ Let's do a thought experiment. A box of equilibrium gas is separated into left and right parts, with a small gate in between them (Fig. 9.1). The gate keeper is a little demon (known as Maxwell's demon). The demon only allow molecules with $v > v_0$ to enter the right part and molecules with $v < v_0$ to enter the left part. After some time, would the right part hotter than the left? If so, is the second law of thermodynamics violated?

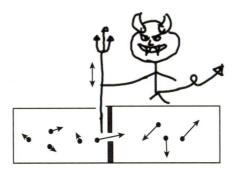

Fig. 9.1 Maxwell's demon

9.1 The Statistical Entropy

To take a closer look at entropy, let us first review the condition of equilibrium:

Equilibrium of subsystems: thermodynamics

Consider an isolated system containing two sub-systems A and B. A and B can exchange heat but each has fixed volume and fixed particle number. The total energy

---
[1] For other such quantities such as internal energy E, volume V, pressure p, temperature T, we can feel them and know what we are talking about.

$E = E_A + E_B$ is fixed since the whole system is isolated. Suppose that A and B are in thermal equilibrium individually.

Under which condition is the whole system in equilibrium?

They have the same temperature. Using the first law Eq.(9.2) (with fixed volume and thus $dV = 0$), we have

$$\frac{dS_A}{dE_A} = \frac{dS_B}{dE_B} \tag{9.4}$$

since both of them equal to $1/T$.

Beyond equal probability

What if we relax our assumption such that each microscopic state has different probability to appear? Then the Boltzmann's entropy formula needs to be generalized. Here we only quote the result:

$$S = k_B \sum_i p_i \ln \frac{1}{p_i} \tag{9.5}$$

where the summation runs over all microscopic states. This formula is known as the Gibbs entropy. One can easily check that if for all i, $p_i = 1/\Omega$, then Gibbs entropy returns to Boltzmann entropy. The Gibbs entropy can be further generalized into the entanglement entropy (von Neumann entropy) in quantum mechanics.

Equilibrium of subsystems: the statistical point of view

Consider the same system as in the previous box. We know the total energy E of the system, and volume and particle number of individual subsystems A and B. Given this knowledge, there are exponentially many possible microscopic states satisfying these constraints (these microscopic states as a collection is known as an ensemble).

We *assume* that each possible microscopic state has equal chance to appear.

What is the nature of the equilibrium state between subsystems A and B? Say, how to partition E into E_A and E_B? The partition should correspond to a maximal number of microscopic states, to maximize its probability to appear.

For given E_A, $E_B = E - E_A$ is fixed. Let's denote the number of possible microscopic states of the subsystem A and B as $\Omega_A(E_A)$ and $\Omega_B(E_B)$, respectively.

The total number of microscopic states is

$$\Omega = \Omega_A(E_A) \times \Omega_B(E_B) \tag{9.6}$$

For Ω to be extremal (most number of microscopic states), we need

$$0 = \frac{d\Omega}{dE_A} = \Omega_B \frac{d\Omega_A}{dE_A} + \Omega_A \frac{d\Omega_B}{dE_B}\frac{dE_B}{dE_A} = \Omega_B \frac{d\Omega_A}{dE_A} - \Omega_A \frac{d\Omega_B}{dE_B} \tag{9.7}$$

Thus the equilibrium condition is

$$\frac{d\ln\Omega_A}{dE_A} = \frac{d\ln\Omega_B}{dE_B} \tag{9.8}$$

Compared with Eq. (9.4), we conclude that

$$S = k_B \ln \Omega \tag{9.9}$$

where k_B is a constant, known as the Boltzmann constant. Detailed comparison between thermodynamics and statistical physics gives $k_B \approx 1.38 \times 10^{-23}$ J/K. This relation is discovered by Boltzmann in the 1870s and Planck put it into the current form in around 1900.

To summarize what we learned here, entropy measures (the logarithm of) the number of possible microscopic states. This can be considered as an indicator of disorder. The 2nd law is thus a more universal saying of "if you do not clean up your room, it will automatically become more disordered instead of cleaner".

Entropy and disorder

To be more intuitive about entropy and disorder, consider a box of gas. Each particle can stay either in the left half or right half of the box. An ordered situation is that all particles stay in the left (or right); and a disordered situation is that half particles stay in the left and half stay in the right. There are much more states for the disordered situation. We leave the details to an exercise.

9.2 The Arrow of Time

There is an arrow of time. Time flows toward the future and do not come back. Who has stolen your time? Spilled water take much more efforts to be gathered up again. Who has stolen your efforts?

Are past and future equivalent?

In Newton's mechanics of a particle, $F = m\dfrac{d^2 x}{dt^2}$. If we perform the time reversal transformation $t \leftrightarrow -t$, we note that the form of the equation does not change its form. (The action of Newton's mechanics also has the time reversal symmetry).

In fact, for all the known fundamental laws of nature, if we take $t \leftrightarrow -t$, and left to right (parity), and particle to anti-particle (charge conjugate), then nothing changes.

In other words, at the level of a fundamental particle, there is no difference between past and future.

However, we do feel the difference between past and future. Why?

Why is there a psychological arrow of time?

What is the psychological difference between past and future? We remember the things in the past, but not the things in the future.

Why we can only remember things in the past? Because getting our brain prepared to remembering things needs increase of entropy (Fig. 9.2). Thus the psychological time arrow has to agree with the thermodynamic time arrow, defined by $\Delta S \geqslant 0$.

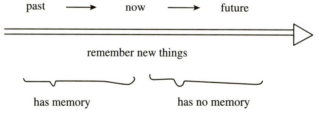

Fig. 9.2

Why $\Delta S \geqslant 0$?

The second law of thermodynamics is somewhat psychological: Even if we live in a universe where entropy decreases, we still feel increase of entropy because we feel that the time arrow is reverted (we can only remember things in the other time direction).

But why not $\Delta S = 0$?

However, according to statistical physics, a system is more likely in the macroscopic state with maximum underlying microscopic states. If it is the case for our universe, then we should stay in equilibrium almost forever and have $\Delta S = 0$. Why is our universe not the case? Unfortunately, we don't know a definite answer at the moment.

Other time arrows in physics

There are a few other physical effects which are not time reversal invariant:

(1) The collapse of wave function after measurement. This is probably because of decoherence and follows the thermodynamics arrow.

(2) Radiation of charge propagates to the future, not the past. This is again related to the thermodynamic arrow of time. The situation is similar to that we often see a stone drops onto water and water waves propagated away. But we do not often see water waves propagates inward and excite a stone out. (Except in carefully designed devices, for example the Chinese fish basin.)

(3) Black hole (classically) absorbs matter but does not emit matter. It is also related to entropy as we will comment below.

(4) Weak interaction. As we mentioned before, we have to revert parity and charge together with time reversal to make weak interaction invariant.

(5) Cosmological expansion. There is no evidence to show that the cosmological expansion is tied to the thermodynamic time arrow. It may be an independent arrow of time.

Entropy and black holes

Why the earth is spherical? High objects will fall (Fig. 9.3).

Fig. 9.3

The black hole has infinitely deep gravitational potential. Thus there is no hill on a black hole. Classically, a black hole contains as little information as a fundamental particle: its mass M, angular momentum L, and charge Q. So at first sight, a black hole carries as little entropy as a particle: $S \sim k_B \times O(1)$.

Then what if we throw something into the black hole? Would the number of possible microscopic states decrease and thus $\Delta S < 0$?

Taking quantum mechanics into consideration (Bekenstein & Hawking), black holes in fact have the hugest number of microscopic states: $S = k_B A/(4l_P^2)$, $l_P \equiv \sqrt{G\hbar/c^3} \simeq 1.6 \times 10^{-35}$ m. For example, for a black hole with horizon area 1cm^2, the information stored is $10^{65} \sim 2^{216}$, which corresponds to 10^{53} hard disks, each with 1TB capacity. Hard disks are made in the unit of nanometer, while black hole entropy is in the unit of Planck length (10^{-35} m).

9.3 Entropy and Information

Maxwell's demon

Recall ③ at the beginning of this part: Can the Maxwell demon reduce entropy? If so, can we use it to do work as a perpetual motion machine?

If the demon indeed *knows* the states of the molecules as Maxwell has described, then indeed, he can use it to reduce entropy and do work. As we know,

> Knowledge is power. — Francis Bacon

Now, does it contradict with the 2nd law? No. Information is not free lunch. There are two possibilities (or combination of them) for the demon to know the states:

(1) The demon has a huge memory in his brain, to record all the molecule data. In this case, the demon's brain must also contain a huge number of possible configurations Ω_D. The entropy of the demon's brain $S_D = k_B \ln \Omega_D$ must be considered as well. The cost of making the gas from disordered to ordered is to make the demon's brain from ordered to disordered.

(2) The demon has little (say, only one bit of) memory. So he has to measure and know the information about the molecules one by one when they arrive. To know the state of one molecule, the demon has to erease the state of the previous molecule from his mind. Landauer (1961) noticed that, to erase one bit of information, the minimal entropy

generated is $k_B \ln 2$ (from the 1st law, the associated heat is at least $k_B T \ln 2$). Adding this part of increased entropy, the 2nd law is not broken.

The Maxwell's demon unveils deep connection between entropy and information. What's information?

From a newspaper editor to an information theorist

Journalists may be among the first people who understand information deeply. They benefit from news (information). What is a piece of news (information)?

For example, in 1909, a newspaper editor Greeley said,

If a dog bites a man that's nothing; but if a man bites a dog, that's news.

Following the spirit of Greeley, what's two pieces of news (information)? Two men bites two dogs, respectively and independently, are two pieces of news.

What can we observe about the nature of information from above?

(1) A man bites a dog is more informative than a dog bits a man: for an event with probability p to happen, the information content $h(p)$ should be a decreasing function of p. For example, $h(p)$ may be related to $1/p$ in some ways.

(2) Two men bites two dogs are two pieces of information: Information is additive for the information content of independent events. For two events with p_1 and p_2 probabilities to happen, the probability for both of them to happen is $p_1 p_2$. Thus the informaiton content should satisfy $h(p_1 p_2) = h(p_1) + h(p_2)$.

The most natural function to satisfy the above observations is [1]

$$h(p) = \log_2 \frac{1}{p} \tag{9.10}$$

The unit of information content is *bit*, since it reflects the minimal binary bits needed to encode the corresponding information.

Thought experiment: efficient cheating in an exam

Let's do a thought experiment, i.e. don't do it in reality: cheat in an exam with your partner, by sending answers of multiple choice questions (one of A, B, C, D) by binary code. You would like to send minimal length of binary codes to minimize the chance of being caught. Suppose you know the probability p_A, p_B, p_C, p_D, how to encode the cheat

---

[1] Here we have used logarithm of base 2 just to match the convention of information theory. It's trivial to convert it to physicists' convention by noting $\log_2 x = \ln x / \ln 2$.

message most efficiently?

Instead of a general discussion, let's consider two examples:

(1) $p_A = p_B = p_C = p_D = 1/4$. The information conetent for each choice is $\log_2(4) = 2$. In other words, each piece of information deserves two binary bits of information. We may encode like $A \to 00$, $B \to 01$, $C \to 10$, $D \to 11$.

(2) $p_A = 1/8$, $p_B = 1/8$, $p_C = 1/2$, $p_D = 1/4$. This is more interesting. A naive cheater may still use two bits to send each choice. But is it the most efficient?

It is intuitive to note that C is more likely to appear, and thus it deserves a shorter code. A and B are rare so we can bear with longer codes. Formally, we can compute: $h(A) = 3$, $h(B) = 3$, $h(C) = 1$, $h(D) = 2$. That suggests that we can encode like $A \to 000$, $B \to 001$, $C \to 1$, $D \to 01$.[①] This is known as the Huffman encoding.

Let us now verify that the Huffman encoding is indeed more efficient. To do so, we compute the *average information content*, i.e. average length of binary code for sending one answer:

$$\langle h \rangle = \sum_{i=A,B,C,D} p_i h(i) = \sum_{i=A,B,C,D} p_i \log_2 \frac{1}{p_i} = 1.75 \qquad (9.11)$$

This is indeed shorter than that of the fixed length coding (2 binary numbers per answer). Shannon's source coding theorem says that you cannot get any better. This is why lossless compression (something like zip, rar, 7z in your computer with smaller file size than original) is possible.

Have you found the equation above similar to something we have learned in physics?

Information entropy

Inspired by the above discussion, the information entropy is defined as the weighted average information content for all possible outcomes. If we have a set of events X, each event $i \in X$ has a probability p_i to happen, then the entropy is

$$H(X) = \sum_{i \in X} p_i \log_2 \frac{1}{p_i} \qquad (9.12)$$

$H(X)$ is known as the Shannon entropy (Shannon 1948). Again, it means the average information content.

Back to physics, up to a constant, the information entropy is nothing but the Gibbs entropy:

$$S = (k_B \ln 2) \times H \qquad (9.13)$$

[①] This encoding can be done easily using a Huffman encoder.

where the p_i in the definition of H is now restricted to physical microscopic states.

9.4 Epilogue: Summary and What's Next

Further reading

(1) For statistical mechanics, find a textbook to find more. For example, *Statistical Mechanics*, Pathria and Beale.

(2) For black hole entropy, read *The Black Hole War*, Susskind.

(3) For information theory, read Shannon's paper: *A Mathematical Theory of Communication*.

(4) For connection between information theory and statistical physics, read *The Physics of Information*, Bais and Farmer.

What happens next?

Statistical mechanics will be the next course to learn for understanding entropy, thermodynamics and much more based on the microscopic states of the system. There is another course of information physics. And definitely information theory is a corner stone of computer science and modern communication. There are many related courses in the School of Engineering.

Exercises

E9.1 Count the number of states in a simple system

Consider gas in a box (classically without involving quantum mechanics) containing N distinguishable molecules. Let us coarse-grain the state of each molecule as being in the left half and the right half of the box. Then each molecule has two states: L or R. Compute the number of states with M molecule in the left half and $N - M$ molecules in the right half. Find the entropy of such state and find the maximal entropy state varying by M.

E9.2 Find the different object using a balance

Given 12 balls, one and only one of them has different weight. With a balance, how to use fewest measures to make sure to find out the different one?

Hint: for the first measurement, one can calculate the information content and entropy of all possibilities and choose the measurement method with maximal entropy. Then the second measurement, etc.

Chapter 10
Complexity

There are great complexities in the science of complexity, since our world contains all kinds of complexities. In this part, we take a glance of the complexity world by a few examples.

Complexities in simple functions

Do you understand a complex function $f(z) = z^2 + C$? Do you understand a real function $g(x) = R(x - x^2)$? If you think you do, take a look at the following images (Fig. 10.1):

Fig. 10.1

The four figures in the left are results of iteratively apply $f(z)$ (at various zoom levels) and the figure in the right is the result of iteratively apply $g(x)$. Did you realize the complexities contained in these functions?

In these figures, there is a hidden number $4.669\,201\,609\cdots$. Can you find it out?

How to measure the length of a coastline?

Let's measure the coastline of Britain by a straight and rigid ruler (Fig. 10.2).[1]

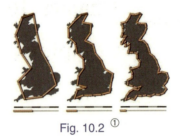

Fig. 10.2 [1]

The length is indicated by the first image to the left.

Let's use a shorter ruler to do the measurement again. Oh, I got a result longer than before. Even shorter ruler? Even longer results. Thanks to the fact that the world is made of atoms, which gives us a short distance cutoff. Otherwise, with infinitely short rulers, I would have measured the coastline of Britain with infinite length!

The Britain has a coastline with infinite length classically? Are you serious?

Four dimensional human beings

From scientific fictions, you may have heard of beings living in 4 dimensional space (excluding time). They are said to appear and disappear suddenly in our view as 3 dimensional beings.

Do you admire them? What if I tell you that we are actually 4 dimensional beings? Can you imagine in which sense this may be real?

10.1 Iteration: from Population of Rabbits to Chaos

The logistic map $R(x - x^2)$ as generations of rabbits

Let's consider models about the population of rabbits on an island after generations of self-reproduction.

In the simplest model, the rabbits has a constant self-reproduction rate R, i.e. after

---

[1] Image from Wikipedia, CC BY-SA 3.0.

a generation, the number of rabbits becomes $n_{k+1} = Rn_k$. The number of rabbits then increase or decrease exponentially by $n_1 = Rn_0$, $n_2 = R^2 n_0$, etc.

This model is too simple. It did not take into account that if there are too many rabbits, the lack of resource on the island will course massive death of the rabbits and thus the population would decrease. To take this into account in a simple toy model, we let

$$n_{k+1} = R(n_k - \frac{n_k^2}{S}) \tag{10.1}$$

where S is a constant. Now the iteration equation appears to have two parameters R and S. An equation with two parameters introduce additional complexity to study. However, without lose of generality, one can redefine $x_k \equiv n_k/S$. Then the iteration equation for x_k is

$$x_{k+1} = R(x_k - x_k^2) \tag{10.2}$$

This stands for iteratively apply a function $g(x) = R(x - x^2)$, and calculate $g(g(g(\ldots g(x_0)\ldots)))$ with an initial value x_0. This function is known as the logistic map.

Behavior of $R(x - x^2)$ with different values of R

How do iterations of $x_{n+1} = R(x_n - x_n^2)$ behave at large n for different R? One can easily write a computer program to test that. And the result for $n \to \infty$ is as follows (Fig. 10.3):[1]

(1) For $0 < R < 1$: $x_\infty = 0$. Because the birth rate is too low to compensate the death.

(2) For $1 < R < 3$: $x_\infty = (R-1)/R$. This value is known as an attractor since it can be obtained from any $0 < x_0 < 1$. To get it, assuming there is a stationary limit, and solve the equation $x_\infty = R(x_\infty - x_\infty^2)$.

(3) For $3 < R < 3.44949$: there is no unique limit x_∞. Instead, at large k, x_k oscillates with two possible values. This can be understood as: for one generation, there are too many rabbits thus many died. The next year, there are more resources and then can support more rabbits.

(4) For $3.44949 < R < 3.54409$: x_k oscillates with 4 possible values. This is hard to understand, but can be tested easily numerically. Further, for $3.54409 < R < 3.56407$: x_k oscillates with 8 possible values; For $3.56407 < R < 3.56876$: x_k oscillates with 16 possible values; ... When one increases R a bit, the period doubles.

(5) For $R > 3.56995$, one can no longer find oscillatory behavior. The behavior of x_k at large k looks random, and is exponentially sensitive to small variation in initial condition

---

[1] Plots of x_k for different values of R. The black solid curve is for $x_0 = 0.2$ and the dashed curve is for $x_0 = 0.205$. The difference is minor for small R but is huge for the chaotic case, indicating a sensitive initial condition dependence.

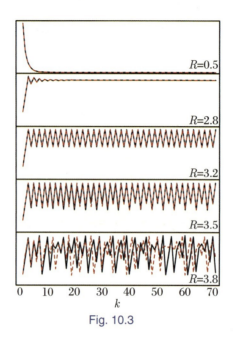

Fig. 10.3

x_0. This behavior is known as *chaos*.

The bifurcation plot

The plot of the logistic map is best known in the form of the below bifurcation plot (Fig. 10.4). Here one clearly note the doubling of periods and eventually the emergence of chaos as R increases.[1]

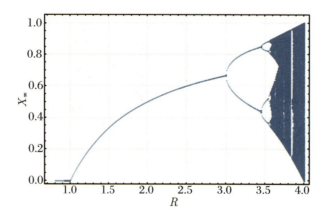

Fig. 10.4

---

[1] If you zoom in the bifurcation plot, you will find some self-similarities in its substructures.

The orderly era: period doubling and the Feigenbaum constant

From the above observations, we note that the oscillation has 2^n possible values for a range of R: $R_n < R < R_{n+1}$. At large n:

$$\delta \equiv \lim_{n \to \infty} \frac{R_{n+1} - R_n}{R_{n+2} - R_{n+1}} = 4.669201609 \cdots \quad (10.3)$$

This number δ is known as the Feigenbaum constant. Interestingly, the Feigenbaum constant is not only the ratio for rabbit birth rate, but rather it is a universal constant of nature. The periodicity of a broad class of non-linear behaviors has the same limiting behavior.

The Mandelbrot set

For the complex function $f(z) = z^2 + C$, the Mondelbrot set is a plot about how fast the sequences $f(0), f(f(0)), \ldots$ diverge or converge. The complexities in this iteration in the complex plane of C is plotted at the beginning of this part. Along the real axis, the plot has self-similarity and the period is determined by the Feigenbaum constant. The relation to the logistic map is plotted in Fig. 10.5.

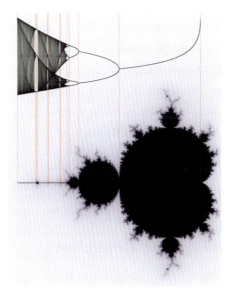

Fig. 10.5

To study the Mandelbrot set in more details, you will find that at different parts of the Mandelbrot set, iterations converge to different number of points. This is very similar to the logistic map. Also note the self similarities in the Mandelbrot set. We will come back to self similarities later.

You can find on the site https://sites.google.com/site/logicedges/logicedges the zoom of the above connection and more videos about complexity.

The chaotic era: the classical loss of determinism

In the framework of classical mechanics, Laplace proposed that if a smart demon knows the initial position and velocity of all particles in the universe, the demon can predict the future. It is known as the Laplace's demon.

Now we know that determinism is (at least apparently) lost in quantum mechanics. Thus, Laplace's proposal does not work. However, even in the framework of classical mechanics, due to chaos, Laplace's proposal does not practically work either. This is because a tiny small error in the knowledge of the initial condition will be amplified exponentially fast in a chaotic system. As a result, it is exponentially hard to predict the future even in the classical sense.

In the 1890s, Poincare studied the 3-body problem and noted that the solutions are exponentially sensitive to initial conditions. This is the starting point of the study of chaos. But for decades, his work was not well understood.

In 1961, Lorenz tried to use 12 nonlinear PDEs to model the change of weather. One day, he wanted to re-run a previous simulation from the middle using data printouts. However, he found that his previous calculation is not reproducible. He struggled and eventually found the reason: the result is exponentially sensitive to the initial condition. This re-discovery is the modern start for chaos study and is later known as the "butterfly effect": "Hurricane formed because a butterfly flapped her wings several weeks earlier."

10.2 Fractals: Dimensions Reloaded

To understand how coastlines would have infinite classical length, let us consider a toy model, known as the Koch's snowflake.

Koch's snowflake

Starting from a straight line, shrink it by 1/3 and fold it 4 times, as Fig. 10.6. Do it *infinitely many times*. Then you get the Koch's snowflake. We observe a few features of the Koch's snowflake:

(1) The snowflake is self-similar. Take a part and zoom in, you find the same image as the original whole image. This non-trivial self-similarity is known as a fractal structure. Though coastlines do not have exact fractal structure, they have fractal-like structures that the part "looks like" the whole after zooming in. You can try to see this feature yourself by zooming in a map application.

(2) What's the length of lines in the snowflake? Let's take the initial straight line to have length L. Then the first iteration of shinking-folding procedure gives a length $(4/3)L$, the second iteration gives $(4/3)^2 L$, ... and the n-th iteration gives $(4/3)^n L$. As $n \to \infty$, the length diverges. Thus, the Koch's snowflake fits an line with infinite length into a finite width (and height).

(3) Due to the folding (roughness), the snowflake appears to have a bit of "thickness" if you look from far away. Thus, the snowflake appears in some sense like a two-dimensional object instead of a one-dimensional line.

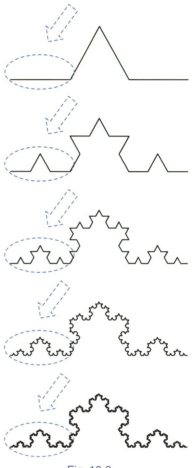

Fig. 10.6

However, if you look infinitely close, the snowflake looks one dimensional again. What's the dimension of the slowflake?

To study the dimension of the Koch's snowflake, let us generalize our definition of spatial dimensions.

The dimension of the snowflake: Hausdorff dimension

How to formally define spatial dimensions? There are many ways. Among which one is by the scaling behavior of objects. Take a square for example, it has trivial self-similarity. Cut its length of edges by 1/2, you get 4 self-similar squares (Fig. 10.7).

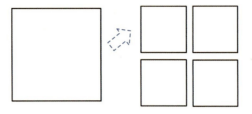

Fig. 10.7

> **The Mandelbrot set**

As another example of fractals, the below image is the Sierpinski triangle (Fig. 10.8).

Fig. 10.8

When you cut edges by $1/2$, you get 3 self-similar pieces. Thus $D_H = \log_2 3/\log_2 2$.

In general, if you cut its length of edges in $1/M$ intervals, and you get N self-similar squares. What's the relation between M and N? Of course, $N = M^2$, and thus

$$\frac{\log_2 N}{\log_2 M} = 2 \tag{10.4}$$

Recall that the square is a two-dimensional object. Is this 2 appearing at the right hand side a coincidence? You can try a 3-dimensional box, or higher dimensional objects. Soon you can be convinced that it's not a coincidence. And in general you have:

$$D_H \equiv (\text{dimension}) = \frac{\log_2 (\text{number of self-similar pieces})}{\log_2 (\text{number of subdivision of edges})} \tag{10.5}$$

Space dimension defined in this way is known as the Hausdorff dimension.

What's the Hausdorff dimension for Koch's snowflake? Cutting the length by $1/3$ and we get 4 self-similar pieces. Thus $D_H = \log_2 4/\log_2 3 \simeq 1.26$. It's an object somewhere between dimension 1 and 2. This is a mathematical statement of the above argument that roughness turns into line width.

You can find a list of beautiful fractals and their Hausdorff dimensions on Wikipedia.

Coastlines and other fractal-like objects

For coastlines, there is no precise self-similarity but at best statistical self-similarity. Nevertheless, one can still count the fractal dimension by scaling properties: Draw grid lines on the map (Fig. 10.9). Count how many boxes cross the coastline. And study how the number of boxes scales when we shrink the intervals between grid lines. One finds that the UK coastline has a box counting dimension of 1.25. The dimensions of coastlines of Norway, Australia and South Africa are 1.52, 1.13 and 1.05, respectively. The smaller dimensions, the smoother the coastlines.

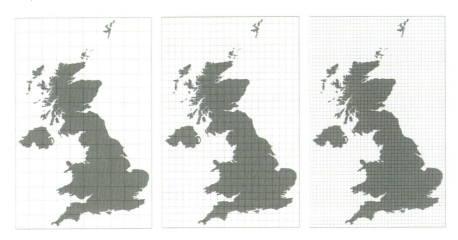

Fig. 10.9

Fractal-like structures are abundant in nature. For example, the fractal dimension of cauliflowers is measured to be about 2.8 (Fig. 10.10). This is intuitive since a cauliflower is a 3-dimensional object and has holes in it arising from fractal structures.

Fig. 10.10

The Kleiber's Law: Metabolic rate for organisms

Let's change topic to a bit of biology. Guess what's the relation between metabolic rate and mass for organisms (Fig. 10.11)? Are they linearly proportional?[①]

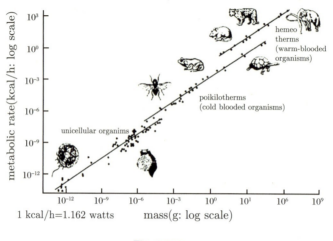

Fig. 10.11

Here is a plot by Hemmingsen on Reports of the Steno Memorial Hospital (1960). Interestingly, the metabolic rate scale as $M^{3/4}$. This strange power is known as the Kleiber's law. The factor 3 in 3/4 is intuitive — this is how mass scales with length. However, where does the factor of 4 come from? Researchers further found that the heart rate and growth rate scales as $-1/4$, and the blood circulation time and life span scales as $1/4$ of animal mass, respectively (Fig. 10.12). Again, the mysterious factor of 4 arises.

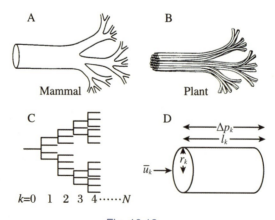

Fig. 10.12

[①] Image from West, Brown, Enquist, Science 276, 122 (1997). See also their paper on Science 284, 1677 (1999).

In the 1990s, West, Brown and Enquist noted that fractal-like structure (such as blood vessels) is more efficient in deliver materials in the body of animals. The fractal-like structure in 3-dimension corresponds to 4 dimensions if no fractal-like structure is used. Thus, in the fractal sense, we effectively live in 4 spatial dimensions!

10.3 Epilogue: Summary and What's Next

Further reading

(1) There is an excellent online course *Introduction to Complexity* by Melanie Mitchell. She also has a book *Complexity: A Guided Tour*.

(2) For more formal textbooks, see, for example, *Fractal Geometry: Mathematical Foundations and Applications* by Falconer.

(3) Many models here can be simulated in a modeling language "NetLogo". The model source code is simple and fun to read thanks to the language features provided by NetLogo.

What's next?

You may like to learn about information, emergent behaviors, networks, genetic algorithms and cellular automata in the world of complexity from the suggested further readings.

Complexity is an interdisciplinary science. You may find your next steps in how to route computer networks, how to recommend goods to customers, how brain works in life science, how birds fly in groups, how to reach a person using your social connections, how to plan the expansion of a city, how to design a algorithm for a robot, how to make more precise weather forecast, and even how a black hole absorbs matter.

Index

α-decay, 129
4-dimesional momentum, 59
4-step reasoning, 17
4-vector, 48

action of a relativistic free particle, 201
action of Newtonian mechanics, 189
action of the Standard Model, 192
action principle, 184, 189
aether, 10
angular frequency, 99
arrow of time: psychological, 224
arrow of time: thermodynamical, 225
asymptotic safty, 211
atom: mean free path, 145
atom: speed of thermal motion, 145
atom:size, 146
atom:spectroscopy, 150
azimuthal quantum number, 156

Bell inequality, 180
bending of light, 71
bifurcation plot, 234

black body radiation, 96
black hole, 75
Bohr's atomic model, 152
bosons, 136
bound state, 130, 132
Brownian motion, 147

Cauchy problem, 183
causality, 32
causality and relativity, 33
chaos, 234
classically allowed region, 127
classically forbidden region, 127
coherent state, 100
collapse of wave function, 118
Compton effect, 101
connection conditions of wave function, 126
conservation laws: importance, 52
conserved quantity, 195

Dalton's multiple proportion law, 143
double-slit experiment, 96
double-slit experiment, single photon, 103

eigenfunction, 116
eigenstate, 116
Einstein's equations, 81
electron cloud, 158
elementary particles, 203
energy, 59
energy band, 134
energy conservation, 196
ensemble, 222
entanglement, 175
entropy: Bekenstein-Hawking, 226
entropy: Boltzmann, 223
entropy: Clausius, 220
entropy: Gibbs, 222
entropy: Shannon, 228
entropy: von Neumann, 222
EPR paradox, 178
equation of motion, 183
equivalence principle, 70
Euler-Lagrange equation, 190
event horizon, 76
events, 11
expectation value, 109

Feigenbaum constant, 235
Fermat's principle, 184
fermions, 136
fluctuation-dissipation theorem, 149
force, 55
frequency, 99
from quantum to classical, 199
functional, 186
functional variation, 186

Galileo's relativity, 3
Gaussian wave packet, 124
geodesic equation, 81
global phase, 112
grand unification, 205
gravitational lensing, 71
gravitational mass, 69
gravitational time dilation, 73
gravitational waves, 77

ground state, 132

Hamiltonian, 196
hidden variable theories, 179
hyperbolic functions, 46

identical particles, 135
inertial mass, 69
information content, 228
invariant interval, 48
invariant mass, 59

Kleiber's Law, 240
Koch's snowflake, 237

ladder paradox: circuit, 38
ladder paradox: closed garage, 37
ladder paradox: open garage, 36
ladder paradox: trap, 37
Lagrangian, 190
Landauer's principle, 226
Langevin equation, 148
length contraction from moving ruler, 24
lepton, 203
light clock, 14
light cone, 34
logistic map, 233
loop quantum gravity, 211
Lorentz transformation, 41
Lorentz transformation and rotation, 46
Lorentz transformation: space, 40
Lorentz transformation: time, 41

magnetic moment, 166
magnetic quantum number, 156
Mandelbrot set, 235
mass: not conserved, 56
matter wave, 106
Maxwell's demon, 221, 226
metric, 49
Michelson-Morley interferometer, 9
microscopic states, 222
Minkowski space, 48
momentum, 54

momentum operator, 115
momentum: why not Newtonian, 52

neutrino, 203
Noether theorem, 195
non-renormalizable theories, 210
null, 35

observable, 116
orbital, 158

path integral, 198
Pauli matrices, 171
Pauli's exclusion principle, 136
period, 99
periodic table, 161
photoelectric effect, 97
photon, 100
Planck energy, 207
Planck energy collider, 212
Planck length, 207
Planck mass, 207
Planck time, 207
plane waves, 108
postulates of special relativity, 11
potential: barrier, 128
potential: double barrier, 133
potential: double well, 133
potential: step, 125
potential: well, 130
potential:periodic, 134
principal quantum number, 157
proper time, 54

quanta, 100
quantization, 100
quantum computing, 175
quantum gravity, 81
quantum gravity: difficulty, 211
quantum gravity: necessity, 206
quantum information, 174
quantum non-cloning theorem, 121
quark, 203
qubit, 169

relative phase, 111
renormalizable theories, 210
resonant tunneling, 133
rest energy, 57
rigid body (non-existing), 34
rotation, 39

same time, 27
Schrödinger equation, 122
Schrödinger's cat, 138
Schwarzschild metric, 74
selection rules, 158
shell, 158
Sierpinski triangle, 238
simultaneity, 27
simultaneity is relative, 29
singularity, 77
space-like, 35
speed of light, 5
speed of light from Maxwell equations, 7
spin, 168
spin states: inner products, 170
spin states: operators for measurements, 172
spin states: projection operators, 172
spin states: state vectors, 170
spin states: basis, 169
stationary state Schrödinger equation, 124
Stern-Gerlach experiment, 166
string: dualities, 217
string: landscape, 217
string theory, 82, 211, 215
subatomic structure, 151
superposition, 111
supersymmetry, 205
symmetry, 193

time (Newton), 13
time dilation, 16
time translation, 193
time-like, 35
transformations, 193
tunneling, 129
twin paradox, 19

twin paradox: revisited, 31

ultraviolet catastrophe, 96
ultraviolet divergence, 209
uncertainty principle, 119
unification, 205

velocity addition, 43
velocity addition: first look, 25

wave function, 109
wave-particle duality, 106
wavelength, 99
wavenumber, 99
which-way experiment, 104